Science and Supernature

Science and Supernature

A Critical Appraisal of Parapsychology

James E. Alcock

Prometheus Books
Buffalo, New York

SCIENCE AND SUPERNATURE. Copyright © 1990 by James E. Alcock. All rights reserved. Printed in the United States of America. No part of this book may be reproduced in any manner whatsoever without written permission, except in the case of brief quotations embodied in critical articles and reviews. Inquiries should be addressed to Prometheus Books, 700 East Amherst Street, Buffalo, New York 14215, 716-837-2475.

5 4 3 2 1

Library of Congress Cataloging-in-Publication Data

Alcock, James E.
    Science and supernature.

    Includes bibliographical references.
    1. Parapsychology and science. I. Title.
BF1045.S33A42   1989   133.8   89-70033
ISBN 0-87975-548-2

Part 1 of the present work appeared in a slightly different version in *Behavioral and Brain Sciences*, vol. 10, no. 4 (December 1987). Part 2 is adapted from the author's "Comprehensive Review of Major Empirical Studies in Parapsychology Involving Random-Event Generators or Remote Viewing," commissioned by the Committee on Techniques for the Enhancement of Human Performance, Commission on Behavioral and Social Sciences and Education, National Research Council, Washington, D.C.

Contents

Introduction   1

Part 1: Parapsychology: Science of the Anomalous or Search for the Soul?

    Introduction   9
    1 What Is Psi?   15
    2 Is Psi "Possible"?   20
    3 If Psi Exists, How Can It Be Detected?   22
    4 Is There Any Substantive Evidence That Psi Exists?   29
    5 Does Parapsychology Follow the Rules of Science?   37
    6 Are the Critics Fair?   43
    7 Is Rapprochement Between Psychology and Parapsychology Possible?   49
    Conclusion   52
    Postscript   54

Part 2: Psi in the Laboratory

    Introduction   81
    1 Parapsychological Research Using Random-Event Generators   89
    2 Remote Viewing   111
    3 Overall Conclusions   126

    Appendices   129

References   177

Introduction

Thus spake Descartes: *Cogito, ergo sum.* "I think, therfore I am." But who am I, or rather *what* am I? What are you? Are we simply fleshy automatons, processing information from the environment and acting on it as we must? Or are we creatures of free will, whose whim is enough to alter what we otherwise tend to view as a deterministic world that follows well-documented natural laws? We think. We feel. We are conscious of our own existence, unlike most if not all of the other living things with which we share the planet. And that consciousness makes us egocentric both as individuals and as a species: Some of us believe in a deity or deities that grant us special privilege over all other living things. Others of us arrogate that privilege for ourselves.

Yet, despite our dominion over the other creatures of land and sea, there are times when we can feel so very insignificant: Has anyone ever stared into the sky on a clear dark night and not felt awestruck? The heavens. Home of the gods. Playing field of astrology. Birthplace of science. It was through observing the heavens that Copernicus reasoned that we and our small planet are not at the center of things after all. With this discovery, modern science began. And as science prospered, many of our most cherished beliefs about ouselves gave way to sobering realizations of our relative unimportance in the universe. It has been said that Darwin and Wallace robbed us of our souls, and that Freud took away our minds. Now, in the closing years of the twentieth century, as science probes deeper and deeper into the workings of our brains, as computer builders threaten us with artificial intelligence that may in some ways ultimately surpass our own, and as the philosophical and emotional power of the traditional religions continues to erode in many quarters, it is not surprising that many people think about life and wonder, in the words of the old song,

"Is that all there is?"

For some, the so-called New Age religions, with their easygoing, narcissistic orientation and their assurance not only that we are immortal but that we create our own reality, provide refuge from the existential void. For others, there is simply the belief or hope that there *is* more to our existence than modern science can measure.

Parapsychology is the attempt to study the independence of the mind from the world of the physical. It is the endeavor to show that there is more to this world than is dreamed of in materialistic philosophy. It is the quest to explain the curious, sometimes frightening experiences people report that seem to lie outside the realm of normal experience and seem to suggest a nonmaterial dimension of our existence.

Yet, as I have discussed elsewhere (Alcock 1981), there is no need to invoke paranormal or supernatural explanations for our unusual experiences, if we bother to study the way our brains work and the ways in which we process information. We should expect, from time to time, to have experiences that *seem* to reflect the operation of extrasensory perception or precognition. That does not by itself mean that all such experiences are *not* paranormal, but only that there are many ways in which our brains can mislead us and make us think that something is extraordinary when it is not.

This book is about parapsychology. It deals in part with much of the best research that parapsychology has produced. However, it is based on a critical evaluation of that research. Other evaluators might have come to different conclusions. Indeed, for each of the two parts of the book, I point to corresponding papers by parapsychologists who examine the same research and draw conclusions that are opposite to mind. It is my hope that the reader who takes the time to digest my arguments, and who makes the effort to follow the sometimes detailed descriptions of the research reports, will come to agree that at this point there has yet to be a persuasive demonstration that anything "paranormal" exists.

The book comprises two lengthy papers that have heretofore been relatively inaccessible. One of them, "Science of the Anomalous or Search for the Soul" appeared in the journal *Behavioral and Brain Sciences* in 1987. It was accompanied by an article written by two leading parapsychologists, K. R. Rao and John Palmer, who presented the "best case" for parapsychology. In the same issue, fifty scholars responded to either or both of these two target articles, and then the authors of each of the target articles responded to these critiques. The interested reader is invited to seek out Rao and Palmer's article and their response to the critics, as well as the critics' responses to their article and to mine.

There may be a good lesson for skeptics in the responses of some of the critics: As I discuss in my response to them, only a very little of my article touches on the belief system or parapsychologists, and only then in the context of why parapsychology persists. Yet a number of critics, including some who share my skepticism, chided me for downplaying parapsychologists' research efforts because of my evaluation of their belief systems. Nothing could have been further from my intention. My mistake, and here is the lesson, at least for me, was in the title I chose: "Science of the Anomalous or Search for the Soul." This title appealed to me at least in part because of the way the alliteration rolls off the tongue. Unfortunately, it created in some readers a particular mental set with regard to my viewpoint, a set that overrode what I actually had to say. So, while the message is important, so obviously is the packaging.

The second half of the book consists of the Background Paper I was commissioned to prepare for the Committee on Techniques for the Enhancement of Human Performance of the National Research Council of the United States (NRC). This paper examines in considerable detail parapsychological research involving either remote viewing or the use of random-event generators. To my knowledge, it provides the only in-depth, experiment-by-experiment analysis of Helmut Schmidt's research that has been published.

A word about the context in which this paper was prepared is important. In 1984, the Army Research Institute of the United States asked the National Academy of Sciences to set up a committee to evaluate a number of techniques that have been claimed to have value in enhancing human performance. This NRC committee comprised fourteen experts, virtually all of them psychologists. Subcommittees were set up to examine among other things the effectiveness of sleep learning, guided imagery, biofeedback, neurolinguistic programming and related techniques, and parapsychology. Professor Ray Hyman, a psychologist at the University of Oregon who had long been considered by most parapsychologists to be a fair-minded critic of parapsychology, was appointed a member of the NRC committee and headed the parapsychology subcommittee.

The interest in parapsychology reflected an initiative within the Army to contemplate the usefulness of psychokinesis for influencing the operation of distant machines and the usefulness of extrasensory perception for viewing distant sites (Swets 1988). Thus, remote viewing experiments and research into the influence of the mind's effects on matter were of primary interest.

It is important to note that the Army Research Institute had previously commissioned parapsychologist John Palmer to provide a thorough evaluation of the best of parapsychological research. Palmer's report (Palmer 1985c), which was received by the Army Research Institute shortly before the formation of the NRC committee, was a lengthy and detailed review of eight areas of parapsychology. In his usual scholarly fashion, Palmer offered not only the best case for parapsychology in each of these areas, but also gave a careful presentation of the arguments critics had made and added further critical commentary of his own. Overall, he concluded that despite the criticisms, the body of research still supports the view that some paranormal, or as Palmer would prefer, "anomalous," processes have been demonstrated to exist.

The Army Research Institute instructed the NRC committee to provide a "second opinion" on the set of experiments that Palmer had already evaluated, and in particular to focus only on a subset of those experiments, the ones involving either remote viewing or the use of random-number generators (RNGs). The former type of experiment involves what used to be called "mental telepathy" or perhaps "clairvoyance." The latter type of experiment involves the attempt to demonstrate that the mind can influence processes that are otherwise considered to be truly random, such as radioactive decay. In order to avoid any charge of bias in the choice of experiments, the subcommittee agreed to restrict their analysis to those *specific* experiments included by Palmer in his survey (Hyman, personal communication).

Thus the NRC committee did not include parapsychologists because the Army Research Institute had already obtained a parapsychologist's evaluation. (Interestingly, Palmer and I rarely differed in our evaluations of the shortcomings of the experiments; we did differ in our conclusions about the implications of the body of research.)

The NRC committee in its report (Druckman and Swets 1988) concluded that there is "no scientific justification from research conducted over a period of 130 years for the existence of parapsychological phenomena" (p. 22). It went on to say: ". . . There is no reason for direct involvement by the Army at this time. We do recommend, however, that research in certain areas be monitored, including work by the Soviets and the best work in the United States" (p. 22).

Rather than being encouraged by that last sentence, the parapsychological community was outraged by the committee's report. Its spokespersons (Palmer, Honorton, and Utts 1989) argued that its conclusion was unwarranted and that the evaluation was biased since "the two principal evaluators of parapsychological research for the committee, Ray Hyman and James Alcock, were publicly committed to a negative position on parapsychology at the time the committee was formed. Both are members of the Executive Council of an organization well known for its zealous crusade against parapsychology" (p. 32).

The controversy did not end there. In 1988, the Office of Technology Assessment (OTA) of the United States Congress organized a "working group" on experimental parapsychology at the instigation of Senator Claiborne Pell, who has an abiding interest in parapsychological research. A one-day workshop was held, which turned out to be an opportunity for parapsychologists and their supporters to take issue with Ray Hyman and me with regard to the conclusions reached by the NRC committee in its report and by me in my Background Paper. Neither Hyman nor I anticipated that this would be the theme of the meeting, which we had been told had been organized to bring skeptics and proponents together to see what common ground could be found. The interested reader is directed to the report of that workshop (Office of Technology Assessment 1989), although it is now my turn to warn of bias, for five of the eight participants have been active either in terms of parapsychological research or advocacy or defending parapsychology from skeptics like Hyman and myself, and another admitted to being essentially a "believer" in the paranormal. Thus it was six against two in the discussions that occurred;

and in my opinion, this imbalance is naturally reflected in the account of the workshop proceedings.

Following the publication of the NRC committee's report, specific criticism was raised about the conclusions that I and the committee had drawn with regard to Robert Jahn's research, and these criticisms were also raised at the OTA workshop. Concern was expressed about the exclusion of Jahn's more recent data; but of course, as mentioned earlier, only the experiments already reviewed by John Palmer were considered. Indeed, Palmer and I shared similar critical views of this research, and *both* Palmer and I had concluded that the overall significance in Jahn's research depended largely on the data from a single subject, who happened to be the first subject who participated in the experiment as well as a primary member of the research team itself. Subsequent to the completion of Palmer's report, in 1986 Jahn and his research team published a report in which data from a few more subjects were included in their database; and at the OTA workshop, Jahn presented data from yet another few additional subjects. If one removes the data of the first subject from this expanded pool, the remaining database then (barely) reaches statistical significance at the .05 level. That barely significant finding contrasts with the very, very high level of significance that is found when the first subject is left in. Thus the notion that the bulk of the significance is due to just one subject still obtains. Moreover, since Jahn has carried out a number of statistical tests on his data, the .05 level of significance is not the appropriate criterion for such post hoc testing. By any of the more acceptable methods for computing the proper levels of significance in simultaneous statistical testing, the results for the subjects with the first one excluded would be considered nonsignificant (Hyman, personal communication, 1989).

* * *

In the pages that follow, the reader will learn of the many problems, both conceptual and methodological, that have plagued parapsychological research since its beginning. Given the weakness of the evidence accumulated over a century, the most interesting question must surely be, Why does the pursuit endure?

PART 1

Parapsychology: Science of the Anomalous Or Search for the Soul?

Introduction

It is curious that, in this age of unprecedented literacy and unceasing scientific and technological progress, many people are prepared to accept that spoons can be bent by the power of the mind alone, that disease can be cured by the laying on of hands, that water can be located by means of a forked willow stick, or that the mind can influence the decay of radioactive substances. It is even more curious when such claims are put forth and defended by people trained in the ways of science.

Most of my readers, I would imagine, have little difficulty dismissing popular occult beliefs in astrology, palmistry, the tarot, or biorhythms. However, those same readers may not be nearly so cavalier about disregarding such supposed "paranormal" (also synonymously referred to as "parapsychological" or "psi") phenomena as extrasensory perception ("ESP") or psychokinesis ("PK"). ESP refers to the supposed ability to obtain knowledge of a target object or of another person's mental activity in the absence of sensory contact, and PK is the putative ability of the mind to influence matter directly. Belief in such phenomena is actually very widespread, not only among members of the general public but also among university students (e.g., Alcock 1981; Gray 1984; Otis and Alcock 1982).

Such belief is no doubt tied, at least in part, to the fact that many people, perhaps even most, have from time to time had direct personal experiences that seemed to be "telepathic" or "precognitive" or "psychokinetic." Indeed, a number of surveys (e.g., Alcock 1981; Evans 1973; Irwin 1985a; McConnell 1977; Sheils and Berg 1977) have found personal experience to be the major reason given by respondents for their belief in paranormal phenomena. This is not surprising: Given their often powerful emotional impact, combined with a lack of understanding about the myriad "normal" ways in which these experiences can come about (e.g., see Alcock 1981; Marks and Kammann 1980; Neher 1980; Reed 1972; Zusne and Jones 1982), it is easy to ascribe paranormal explanations to odd experiences that one cannot readily explain otherwise.

"Parapsychology" is defined as the scientific study of paranormal phenomena (Thalbourne 1982). The study of the paranormal was historically associated with the so-called occult sciences, such as astrology and numerology; a more direct progenitor was the spiritualism craze of the late-nineteenth and early-twentieth centuries. However parapsychology stands well apart from these belief systems in a number of ways:

Scientific Orientation

For more than a century, there has been careful and deliberate investigation of psi phenomena by people trained in the methods of science. In the past fifty years, much of this research has been laboratory-based and carried out in university settings. Currently, parapsychological research is being conducted at such prestigious academic institutions as the University of Edinburgh and Princeton University.

Throughout the last century and continuing to the present, a number of very prominent natural and social scientists have been proponents and supporters of parapsychological research (see Hyman 1985a; Rogo 1986), including physicists Sir William Crookes, Lord Rayleigh (Nobel Prize, 1904), Wolfgang Pauli (Nobel Prize, 1945), Brian Josephson (Nobel Prize, 1973), and David Bohm; naturalist Alfred

Russel Wallace; chemist Robert Hare; physiologist Charles Richet (Nobel Prize, 1913); psychologists William James, William McDougall, Carl Jung, Sir Cyril Burt, and Hans Eysenck; anthropologist Margaret Mead; mathematician John Taylor (who became convinced of the reality of psi phenomena on the basis of Uri Geller's purported feats [Taylor 1975], only subsequently to repudiate his belief in such phenomena [Taylor and Balanovski 1979]; and Robert Jahn of the Engineering Department at Princeton University.

There has also been a history of professional interaction between conventional science and parapsychology at scientific conferences, through symposia on the paranormal and invited addresses by parapsychologists (e.g., American Association for the Advancement of Science, 1975, 1978, 1984; American Physical Society, 1979; American Psychological Association, 1966, 1967, 1975, 1984, 1985), although admittedly such opportunities for parapsychologists to present their ideas and evidence have been limited.

Organization

As a research discipline, parapsychology is organized very much the way various disciplines of mainstream science are. There are professional bodies that emphasize empirical inquiry using scientific methodology and that encourage high research standards. (One of these, the Parapsychological Association, more than half of whose 300 or so members hold doctorates in science, engineering, or medicine [McConnell 1983], has been affiliated with the American Association for the Advancement of Science since 1969.) Annual research conferences are held. Research grants are awarded. There is substantial empirical literature in the field, including several research journals and many books, some of which have been published by leading scientific publishers (e.g., the *Handbook of Parapsychology* [Wolman 1977a], the *Foundations of Parapsychology* [Edge et al. 1986], and the series *Advances in Parapsychological Research* [Krippner 1977; 1978a; 1982a; 1984]).

Academic Involvement

Courses in parapsychology are offered for academic credit at about fifty colleges and universities in the United States (McConnell 1983); a few even grant degrees in the subject (see Stanford 1978). Ph.D.'s have been awarded for parapsychological research at Cambridge University, the University of Edinburgh, Surrey University, Purdue University, the University of the Witwatersrand, and the City University of New York, among others. The University of Edinburgh has recently established the Koestler Chair in Parapsychology, which is endowed by a bequest from the late Arthur Koestler, a longtime supporter of parapsychology.

How do members of academia view claims about psi? In one survey of humanities and science professors at two large universities (University of Michigan and University of Toronto; response rate 53 percent), only about one-third of the respondents indicated believing in paranormal phenomena (Otis and Alcock 1982); there was no clear difference between representatives of the sciences and the humanities. this is consistent with the results of a smaller survey conducted at two other Canadian universities (Alcock 1981). Yet, Wagner and Monnet (1979), in a much larger survey of professors at 120 colleges and universities in the United States (response rate 49.5 percent), found that 73 percent of the respondents from the humanities, arts, and education indicated they believed ESP to be either an established fact or a likely possibility, whereas only 55 percent of the respondents from the natural sciences and 34 percent of the psychologists did likewise. Whether the differences between the results of the two surveys reflect differences in the questions asked or differences in the groups sampled (the former study was limited to respondents from two large and prestigious universities) is not clear. (It should be noted in any case that such surveys are always subject to a response bias, in that there is likely to be a differential response rate as a function of attitude toward the subject matter being addressed.)

Although all of this might suggest that parapsychology is a serious and professional research discipline that is viewed with respect within university settings, at best parapsychology struggles to maintain a toehold at the fringes of academia; mainstream science continues virtually to ignore its subject matter or even to reject and ridicule it. One finds no mention of psi phenomena in textbooks of physics or chemistry or biology. Lecturers do not address the paranormal in undergraduate or graduate science programs. Psychology students are rarely taught anything about the subject. Parapsychological research papers are only very infrequently published in the journals of "normal" science, and parapsychologists have criticized leading scientific publications, such as *Science*, the *American Journal of Physics*, and *American Psychologist*, for suppressing the dissemination of parapsychological research findings (Honorton 1978a; McConnell 1983). Funds for parapsychological research are usually generated within parapsychology itself or come from private donors; the agencies that fund normal science turn a blind, or even hostile, eye toward parapsychological research proposals. The United States government, however, has provided multi-million-dollar support for psi research into remote viewing at SRI International in California [Targ and Harary 1984].)

What accounts for the disparity between what would seem to be a substantial degree of professionalism in parapsychology on the one hand, and the continuing relegation of parapsychology to the fringes of science on the other? For one thing, parapsychology continually encounters opposition from mainstream psychology; psychologists appear to constitute the most skeptical group concerning whether psi is likely to exist (Alcock 1981; Wagner and Monnet 1979). Second, people who may serve as the "gatekeepers" of science, in that they are very influential in determining what is and is not the proper subject matter of science, are skeptical about psi. A recent survey of "elite" scientists (Council members and selected section committee members of the American Association for the Advancement of Science) revealed the highest level of skepticism regarding ESP of any group surveyed in the past 20 years (McClenon 1982): Fewer than 4 percent of the 339 respondents (the response rate was 71 percent) viewed ESP as scientifically established. (However, another 25 percent considered it to be a likely possibility, indicating about the same proportion of favorableness as reported by Otis and Alcock [1982], cited above.) Fifty percent considered ESP to be impossible or a remote possibility.

In McClenon's (1982) view, this negativity is based on the threat that paranormal phenomena, were they to exist, would pose to the prevailing scientific worldview. A rather different viewpoint, which is part of the thesis of this paper, is that parapsychology, over its century or so of existence as an empirical research endeavor, has simply failed to produce evidence worthy of scientific status. Of course *both* these views could be correct.

To facilitate the discussion of this issue, I shall proceed by posing a number of questions I consider to be important concerning psi and parapsychology:

1. What is psi; how is it defined?
2. Is psi "possible"?
3. If psi exists, how can it be detected?
4. What is the evidence that psi exists?
5. Does parapsychology follow the rules of science?
6. Are the critics fair?
7. Is rapprochement possible between psychology and parapsychology?

Let us consider each of these questions in turn.

1
What Is Psi?

Although it may at first seem straightforward to define or catalogue paranormal phenomena, it turns out to be a difficult task indeed, for there is a considerable spectrum of opinion even within parapsychology as to which ostensible phenomena are likely to be genuinely paranormal and which are probably based on error and self-delusion. For example, although many parapsychologists might scoff at such claims, some believe that "psychic healers," through the laying on of hands, can speed the healing of wounds and slow the growth of fungi (Krippner 1982b); others believe that some gifted persons can project images onto photographic film (Eisenbud 1977), that water sources, or even lost treasure, can be located by "dowsing" with a willow stick (Bird 1977; Schmeidler 1977), that reincarnation warrants serious investigation (Child 1984; Stevenson 1977), that one's personality can leave and return to the body at will and may even be able to travel through outer space (Targ and Puthoff 1977), and that deathbed visions may be indicative of survival after death (Osis and Haraldsson 1978).

Because there is no general agreement on what psi is, or at least how it manifests itself, parapsychologists have found it easier to define it in terms of what it is *not*. The term *psi* itself was introduced by Thouless (1942) as a neutral label in order to avoid the many associations that terms such as *psychic phenomena* and *extrasensory perception* have developed over the years, and psi is defined simply as "interactions between organisms and their environment (including other organisms) which are *not* mediated by recognized sensorimotor functions" (Krippner 1977, p. 2; my italics). Psi phenomena, then, are explicitly defined in a negative manner: To demonstrate that psi has occurred, one must first eliminate all *normal* sensorimotor explanations.

Although only a few parapsychologists appear to share his conservatism, Palmer (1985a; 1986a) argues that until parapsychologists have produced a positive theory of psi that describes the properties that must be present in order to claim that psi has occurred, all they can claim to have demonstrated is the occurrence of a number of anomalies that themselves constitute the subject matter of psi. Seemingly paranormal events might be explicable in terms of conventional science or science as it will be understood in the future, he says, or, indeed, such events might be due to errors in interpretation or measurement or statistical analysis. He recommends that the term *paranormal phenomena* be supplanted by a much less committed term such as *ostensible psychic events*.

Palmer's circumspection is commendable and would find favor with most critics of parapsychology. However, it is rare to find parapsychological research reports or other kinds of literature treating apparent anomalies in such a noncommittal fashion. Most, in fact, treat psi not as a description of an anomaly but as a causative agent.

There is a second and more important sense in which psi is negatively defined, albeit implicitly, and that is in terms of its incompatibility with the prevailing scientific worldview (Boring 1966; Flew 1980; Mackenzie and Mackenzie 1980): In some way or another, psi phenomena, to be considered as such, are impossible if the current worldview is correct. There are two different camps within modern parapsychology regarding this incompatibility (Beloff 1977):

Incompleteness of Current Science

Just as the scientific worldview changed to accept the extraterrestrial source of meteorites and the constancy of the speed of light, so too, according to this viewpoint, it must ultimately accommodate psi. Thus "paranormal" phenomena are part of the natural order, but a part of that order that is not yet understood; as soon as scientific knowledge advances to the point that the paranormal is comprehensible, then the latter will become part of an expanded normal science (Truzzi 1982).

This process has been manifested already in several instances: Bat navigation was taken to involve psi until the echo-sounding apparatus of bats was discovered, at which time it became part of the normal scientific domain and of no further interest to parapsychologists (Boring 1966). Bird navigation (Pratt 1953; 1956) and hypnosis (see McConnell 1983; Spanos 1986) are other examples of phenomena that have passed from the realm of the paranormal to the normal.

A Nonphysical Dimension of Existence

According to this perspective, paranormal phenomena mark the outer limits of the scientific worldview, and beyond those limits "lies the domain of mind liberated from its dependence on the brain. On this view, parapsychology, using the methods of science, becomes a vindication of the essentially spiritual nature of man which must forever defy strict scientific analysis" (Beloff 1977, p. 21).

Of these two perspectives, the incompleteness approach would no doubt be more acceptable to most scientists. Yet it does not really capture the *flavor* of the paranormal. Whereas anomaly is, it would seem, a necessary condition for paranormality, it is not a sufficient one. Were it sufficient, then all anomalies throughout the history of science would have to have been considered "paranormal," whereas it is clear that they have not been considered as such (Braude 1978).

Braude (1978) suggests that a definition of the paranormal must go beyond anomaly to include the notion that it "thwarts our familiar expectations about what sorts of things can happen to the sorts of objects involved" (p. 241). Yet, as Mabbett (1982) points out in response to Braude, experimental parapsychological studies that are taken to demonstrate the reality of psi typically produce scoring rates that are only slightly above chance; these hardly thwart people's expectations, and "even the thoughtful layman would be unwilling to regard such results as evidence of anything but luck without a little assurance or instruction from the expert statistician" (p. 340).

On the other hand, the bizarre and paradoxical properties of light, as described by relativity theory, would no doubt have been unexpected by laymen as well as by scientists prior to Einstein, Mabbett says, yet most people would not have regarded these properties as paranormal. Mabbett argues that paranormal phenomena are psychological in the sense that they involve mind or consciousness, whatever these may be, and that they reflect a relationship between the mental and physical worlds that is radically different from that conceived of by science.

What is being struggled with here by Braude and Mabbett is that, more than being simple anomalies, paranormal phenomena have a special and particular relationship to the human mind. Indeed, as I have discussed in greater detail elsewhere (Alcock 1985), it is hard to escape the conclusions that the concept of paranormality implicitly involves *mind-body dualism* (see Wolman 1977b), the idea that mental processes cannot be reduced to physical processes and that the mind, or part of it, is nonphysical in nature.

The late Gardner Murphy (1961), once president of the American Psychological Association and one of parapsychology's most erudite and persuasive proponents, argued that even if the paranormal were to be defined only in terms of anomaly, this would still lead to a dualism of some sort because of its independence from considerations of time and space. Indeed, parapsychologists have at times insisted that psi phenomena are distinguished from the other phenomena of psychology by virtue of the fact that they are of a nonphysical nature (e.g., J. B. Rhine and Pratt 1957). Although the boldness of such a declaration might well raise the hackles of some modern parapsychologists, most of them do seem to accept such dualism (Thalbourne 1984). The influence of dualistic thinking creates a deep schism between parapsychology and modern science.

In summary, then, although some modern parapsychologists prefer to speak only of anomalies, these anomalies, if they are to be of continuing interest to parapsychology, must ultimately involve some relationship between consciousness and the physical world radically different from that held to be possible by contemporary science. Some parapsychologists might deny being mind-body dualists, but they would do well to consider just how they are going to define their subject matter without some reference to the independence of the mind from the materialistic realm (L. E. Rhine 1967).

2
Is Psi "Possible"?

Psi phenomena are defined implicitly in terms of their incompatibility with the contemporary scientific worldview. Although many parapsychologists (e.g., Rao 1983) believe that only a major revolution in scientific thought could lead to the accommodation of psi, there have been attempts to reconcile such phenomena with modern science. For example, although it would seem that psi cannot occur without violating well-tested laws of physics—such as the law of conservation of matter and energy and the inverse square law of energy propagation (signal strength is proportional to the inverse square of the distance)— or violating the logical principle that an effect cannot precede its cause, ad hoc explanations of how psi might occur without such violation have been proposed (Collins and Pinch 1982). As an example, with regard to the presumed impossibility of seeing into the future, one could posit that what appears to be precognition is really psychokinesis: The individual uses PK to *bring about* the events he believes have been foreseen precognitively. In a similar fashion, one may be able to construct other *ad hoc* explanations to overcome all the various incompatibilities that appear to exist between physical science and parapsychology, although such contrived mechanisms are not likely to satisfy most scientists.

A more direct attempt to render psi compatible with contemporary science has been made through efforts to show that such phenomena are *not* inconsistent with quantum mechanics. In recent years, there has been considerable discussion in parapsycholgy, led by paraphysicists (parapsychologically oriented physicists) and philosophers, about some of the paradoxes of quantum mechanics and about how it is possible to suggest solutions to these paradoxes that imply the direct influence of the mind on matter, allowing for—or even *demanding*—psi (e.g., Oteri 1975; Schmidt 1975; Walker 1974; 1975).

This has generated negative reaction even within parapsychology (e.g., Braude 1979a), with some paraphysicists, such as Phillips (1979; 1984), arguing that the orthodox view of quantum mechanics does *not* lead to paradoxes that necessitate the introduction of mental influences. Phillips describes the difficulty and the arbitrariness of interpreting the mathematical picture served up by quantum theory: "The predictions of quantum mechanics have been verfied, and there is little doubt that the mathematical formalism is correct. Constructing a physical picture to correspond to the mathematics is much more difficult, and authors differ in what they find intuitively appealing and philosophically satisfactory" (1984, p. 298).

Even if quantum mechanics did allow for psi—a notion few mainstream scientists would be likely to accept at present—that would not in itself make the reality of psi any more likely. Flying cows are not inconsistent with quantum mechanical notions, but as far as we know, they do not exist. What is missing in such discussions of psi is the phenomenon itself. Until there is clear evidence that psi exists, it is surely premature to try to bend quantum mechanics to accommodate it.

3
If Psi Exists, How Can It Be Detected?

There are three major sources of evidence for psi: (*a*) anecdotes of spontaneous personal experiences, (*b*) demonstrations by "gifted" psychics, and (*c*) laboratory experiments. The early studies of psi examined anecdotal reports in great detail, but gradually the realization grew that such evidence is just too unrealiable to serve as data for science (Hövelmann and Krippner 1986; L. E. Rhine 1977; Rush 1986a).

"Gifted" psychics have provided the most spectacular psi claims, both in the early days of psi research and more recently (Rush 1986b). For some parapsychologists (e.g., Beloff 1985), such demonstrations still stand as strong testimony to the reality of the paranormal. Yet, once again, this evidence is unsatisfactory in the extreme, because of both the history of fraud involving reputedly gifted psychics (e.g., see Girden 1978) and, more important, the fact that such psychics have as yet been unable to perform their feats under controlled conditions for neutral or skeptical investigators. For example, Uri Geller was taken by a number of parapsychologists (e.g., Beloff 1975; W. E. Cox 1976; Eisenbud 1976; Hasted 1976; Moss 1976; Puthoff and

Targ 1974) to have genuine paranormal powers until a conjurer's investigations (Randi 1975) showed to most people's satisfaction that Geller was using trickery.

Some parapsychologists (e.g., Schmeidler 1984) insist that the fact that a psychic is caught cheating does not weaken the evidential value of those demonstrations during which the same psychic was *not* caught cheating. Given the rarity of such supposedly gifted individuals, it is not surprising that investigators are loath to terminate their research with an individual just because fraud has been detected on some occasions. However, it is no easy task to guard against fraud if a subject is determined to cheat, and what better indication is there of such determination than the subject's being caught at it?

It was because of dissatisfaction with both anecdotal evidence and uncontrolled demonstrations that Joseph Banks Rhine, in the 1930s, set up an experimental laboratory for the study of psi. The hope was that through rigorous application of the methodology of science, psi would soon be put on a solid empirical footing. Rather than simply relying on the ability of self-proclaimed psychics to demonstrate their skills, Rhine began the systematic study of both gifted and ordinary individuals in a number of "guessing" tasks in which probabilities of success could be calculated. If one makes a prediction, based on a probability model, as to how well a subject should score in a guessing task, or if one predicts the distribution of events whose occurrence depends on a random process (in Rhine's day, dice-throwing; nowadays, subatomic particle emission) that the subject attempts to influence mentally, then if all known normal forces have been ruled out, statistically significant departures from the prediction are taken to indicate the involvement of a psi process. Thus, experimental parapsychology, just as conventional psychology had done before it, took on a pronounced statistical flavor.

If one could reliably demonstrate departures from some statistical model, this would call out for explications. There would be no justification, however, for beginning with an explanation based on parapsychological concepts. If there were unobserved weaknesses in the controls, if some unknown process were involved (e.g., the use of some code based on silent counting, or the use of "silent" dog whistles that children, but not adults, can hear [Scott and Goldney 1960]), if there were equipment problems or biases in the random generator, if the statistical model were inappropriate, or if errors were made in the recording or analysis of the data, the paranormal explanation would be erroneous. Just as important, in the absence of a positive theory of psi, even if an observed effect is not due to artifact, one is left only with an anomaly. The availability of the psi hypothesis can distract the researcher from other, normal explanations and thus impede the development of the understanding of anomalies (Blackmore 1983a).

What would constitute "solid" evidence of psi? Obviously, no evidence is ever 100 percent solid, because we can never be sure how new discoveries will change our understanding of processes that we currently think we understand. Furthermore, evidence that seems unconvincing or unimportant in the light of one theoretical worldview may be viewed as much more important if the prevailing theory changes.

An extraordinary degree of evidence is often demanded in support of extraordinary claims. We are generally less demanding of evidence in the case of claims that "fit" with existing theory or knowledge. When one is weighing evidence in law, the distinction is made between "beyond all reasonable doubt" and "on the balance of probabilities." The former, applied in criminal cases, demands virtual certainty of guilt; the latter, used in civil litigation, refers to the notion that the defendant is more likely than not to be guilty. Because psi is a concept that would probably revolutionize science (Rao 1983), most skeptics implicitly use the criterion of beyond all reasonable doubt, while accepting conclusions made on the balance of probabilities where only "normal" and noncontroversial phenomena are involved. However, although the controversial nature of psi may justify the use of tougher criteria, this view has been attacked as being another tactic for denying legitimacy to controversial claims (McClenon 1984).

Before we accept that psi (even in the simplest sense as an anomaly) has been demonstrated in the laboratory, three important factors must be considered:

Internal Validity

Psychologists use the term "internal validity" to refer to the degree to which experiments are free of the influence of extraneous variables that might introduce alternative explanations for the observed results (Berkowitz 1986). Most criticisms of experimental studies of psi concern internal validity: Randomization may be inadequate, sensory leakage (i.e., communication of information by normal sensory means) may have occurred, and so forth.

McClenon (1984) argues that such methodological criticisms of psi experiments are often unfair. By refusing to accept the shared assumptions that are implicit in any experiment, he says, the critic will sooner or later "ask for information that is no longer available, or for a degree of experimental control and exactitude that is desirable in principle but impossible in practice" (p. 89). Thus, the "perfect" ESP experiment is an impossibility, McClenon contends, for one can always suggest that the experimenter was incompetent or that trickery was involved (see also Honorton 1981). Despite McClenon's concerns, there is a considerable difference between making unsubstantiated charges of incompetence or trickery and pointing to methodological flaws. If the flaws are there, parapsychologists should run the experiments again—without the flaws—rather than argue about the motivation of the person who pointed them out.

Instead of rerunning the experiments correctly, a more usual response is to attack the critic. For example, critics have been chastised for pointing to flaws without demonstrating that these flaws are capable of generating the observed departures from chance (Honorton 1975; 1979; Palmer 1986a). This criticism does not stand up, for two reasons. First, critics are usually not advocating the acceptance of an alternate hypothesis but asking only that claims of psi be suspended until properly controlled studies are carried out (Akers 1984; Hyman 1981). Second, such flaws need not be the *cause* of the statistical deviations, but they *are* symptomatic of law research standards (Hyman 1985b). One should hardly have confidence in the experimental controls if one is faced with evidence of violations of proper procedure. Akers (1984) uses the "dirty test-tube" analogy: A chemist would have little confidence in a colleague's findings if it was observed that a test tube used in the experiment was contaminated.

It is not so difficult to design and execute an experiment that is methodologically and statistically sound. Psychological experiments published in the better psychology journals stand in evidence of this.

Consistency

Before accepting the reality of a phenomenon, one generally looks for signs that there is a consistent pattern of results across experiments. The lack of any consistent pattern in the research findings is one of the most serious weaknesses in the evidence offered for psi (Blackmore 1983a). Unfortunately, it is standard practice in parapsychology to take one pattern of data as evidence for psi in one experiment, then to disregard its absence and take some other pattern as evidence for psi in another experiment.

Repeatability

Not only should there be consistency in the pattern of data across experiments, but individual experiments should be repeatable by others. Repeatability is an important safeguard, albeit only a partial one, against error or fraud (Sommer and Sommer 1984). Obviously, however, replication by itself is not enough. If someone is dishonest in the actual reporting of the research, reports of replication by the same author will not eliminate the dishonesty (Casrud 1984).

Yet, as Rao (1985) points out, repeatability is not a matter of primary concern in normal science. Only if some important and controversial finding is made is replication likely to be attempted, and this will often be undertaken by others who have competing theories that would not accommodate the finding. When observations are consistent with theory, replication is less important. However, as Murphy (1971) commented: "If the event is unclassifiable, then it is doubly important that it have a rational interpretation, that is, one that fits with the thought patterns of the contemporary human mind. If it has no clear rationality, its only chance of demanding scientific attention is replication" (p. 4).

On this basis, repeatability is, in general, less important in psychology than in parapsychology. Even so, psychologists pay far too little attention to the importance of repeatability (Epstein 1980; Fishman and Neigher 1982; Furchtgott 1984; Heskin 1984; Sommer and Sommer 1983; 1984); replication studies account for a very small percentage (3 percent or less) in leading psychology journals (Bozarth and Roberts 1972; Sterling 1959). This has led on occasion to the widespread dissemination of information that is subsequently found to be unreplicable (see, for example, Marshall and Zimbardo 1979; Maslach 1979; Schachter and Singer 1979).

Even when replication is attempted, its importance often depends on who conducts it. We are not likely to accept a wild claim supported by the research of only one person, whether that research has been replicated by that person or not (Hyman 1977a). Similarly, a failure to replicate by a student in a high-school science class will carry little or no weight, whereas a failure to replicate by a well-respected scientist will be much more seriously viewed (Collins 1976). It is also difficult to know just what constitutes a replication (Edge and Morris 1986); there are in fact several different kinds of replication that one can provide (Alcock 1981; Lykken 1968). Beloff (1984) differentiates between "weak" and "strong" replicability, where the former term refers to a situation in which an experiment or phenomenon has been independently confirmed by at least one other investigator, and the latter refers to a situation in which any competent researcher, following the prescribed procedure, can obtain the reported effect. Although parapsychologists have presented, as evidence for psi, studies that have been replicated by other parapsychologists, there has never been a psi demonstration that is replicable in the strong sense (Beloff 1973; 1984; Palmer 1985b). Indeed, parapsychologist/psychologist Susan Blackmore (1983a) recently referred to unrepeatability as parapsychology's *only* finding.

Of course, even if a psi experiment is replicated, that by itself does not mean the effect has a paranormal cause. On the other hand, the inability to repeat an experiment or a demonstration cannot by itself rule out the truth of the psi claim. Poor repeatability could conceivably stem from factors other than the nonexistence of psi (Palmer 1986b). It is possible that certain conditions are necessary for the production of psi; and given that no one knows just what these conditions are, it could be that an essential element is missing when an experiment fails to replicate. It has also been suggested that psi could turn out to be inherently unlawful (Palmer 1986b; Rao 1982), although this position is difficult to defend (Hövelmann and Krippner 1986). From this viewpoint, it has been argued that the quest for repeatability should be abandoned (Pratt 1974).

Despite the arguments about the relative unimportance of repeatability, the history of science demonstrates that unrepeatable experiments or demonstrations should be viewed with a very cautious eye. Most parapsychologists probably would not dispute this point. Indeed, the claim is made that the level of repeatability that has been demonstrated in parapsychology exceeds typical replicability rates in the social sciences; the strongest claim in this regard concerns the psi ganzfeld effect, for which replicability is said to be in the area of 50 percent (Honorton 1976; 1978b). This research is discussed in the next section.

In summary, then, although one cannot set precise standards that evidence of psi must meet, judgment should be suspended until there is at least some consistency among research findings from a body of methodologically irreproachable experiments, at least some of which are repeatable in Beloff's (1984) strong sense.

4
Is There Any Substantive Evidence That Psi Exists?

Within parapsychology itself, there are arguments about the strength of the evidence adduced for psi. Some argue that no substantive evidence has yet been found (e.g., Parker 1978), whereas others consider the laboratory evidence for psi convincing (e.g., Schmeidler 1984); still others believe that psi can even now be harnessed—for example, to guide stock market investments (Targ and Harary 1984). On the whole, it would appear that most parapsychologists believe that psi has already been demonstrated. Schmeidler (1971) reported that almost 90 percent of her small sample of members of the Parapsychological Association indicated they believed that ESP had been established so firmly that any further research aimed only at demonstrating its existence would be uninteresting. Subsequently, in a survey of all 241 members and associates of the Parapsychological Association (which yielded a response rate of 84 percent), 68 percent indicted complete belief in the reality of psi (McConnell and Clark 1980). The average strength of belief across all respondents was 93 percent.

Many studies have been carried out and published that purport to provide statistical evidence for paranormal processes. However, even if we were willing to treat certain statistical deviations as evidence of psi, such evidence has been unsatisfactory: A number of recent analyses have demonstrated a serious problem with the quality of the methodology used in parapsychological research. For example, Akers (1984) cites a survey of 214 PK experiments (May et al. 1980), in which the authors concluded that none had been properly designed and reported.

In order to explore in more detail the state of the evidence in parapsychology, five major areas of contemporary parapsychological research will be discussed below.

Out-of-body Experiences

Blackmore (1982; 1984), after carefully studying both the anecdotal and research literature on out-of-body experiences (experiences in which the individual believes that the physical body has been left behind and that travel through physical space is therefore unencumbered by limitations imposed by the flesh) and after conducting her own research, came to the conclusion that normal psychological theories are capable of accounting for such experiences and that nothing paranormal is likely to be going on.

Personality/Attitudinal Variables and Psi

Akers (1984) evaluated 54 experiments that studied the influence of altered states and of personality/attitudinal variables on psi and that had been cited as significant confirmations of psi. He found that 85 percent of the experiments were seriously flawed, and even the 8 that were conducted with reasonable care were not methodologically ideal. The problems fell into several categories, including randomization failures, sensory leakage, inadequate safeguards against subject cheating, the possibility of errors in the recording of the data, errors in statistical analysis, and failures to report important procedural details.

Akers concluded that these 54 experiments taken together were too weak to establish the existence of a paranormal phenomenon.

The Psi Ganzfeld Effect

As mentioned earlier, studies of ESP in a ganzfeld (a condition of reduced sensory stimulation typically produced by covering a subject's eyes with halved Ping-Pong balls and shining a white light onto them while playing white noise into the subject's earphones) have been very promising in that they have appeared to demonstrate a replication rate of 50 percent or higher (Blackmore 1980; Honorton 1978b).

Hyman (1985b) has completed an exhaustive analysis of virtually all psi ganzfeld research, using a database of 42 studies conducted between 1974 and 1981. Hyman's analysis leads him to conclude that the replicaton rate exhibited in this collection of studies is probably very close to what would be expected by chance. Several flaws of procedure—including less than adequate randomization, the possibility of sensory leakage, and erroneous statistical analysis—plagued these studies; not a single study was flawless, he reported. He suspects that most of these studies were not well planned, and he concludes that this database is too weak to support any assertions about the existence of psi. However, Honorton (1985) disputes Hyman's conclusions, arguing that his assignment of flaws is itself seriously flawed, and he maintains that these studies do indeed indicate a significant psi ganzfeld effect.

Hyman and Honorton (1986) prepared a joint paper as a follow-up to the two papers discussed above. With reference to the database discussed earlier, they agree that "the experiments as a group departed from ideal standards on aspects such as multiple testing, randomization of targets, controlling for sensory leakage, application of statistical tests, and documentation. Although we probably still differ about the extent and seriousness of these departures, we agree that future psi ganzfeld experiments should be conducted in accordance with these ideals" (p. 353).

They go on to say that "whereas we continue to differ over the degree to which the current ganzfeld data base contributes evidence for psi, we agree that the final verdict awaits the outcome of future psi ganzfeld experiments—ones conducted by a broader range of investigators and according to more stringent standards" (pp. 352–353).

Thus, although the ganzfeld studies have been offered as the strongest evidence for a repeatable psi effect, any conclusion about a psi ganzfeld effect must await future research carried out more carefully than these studies were.

Remote-viewing Studies

In 1974 *Nature* carried an article by two physicists (Targ and Puthoff 1974) in which they described their successful demonstrations of "remote viewing," a talent by means of which subjects are able to describe geographical locations being visited by other people without having any normal form of communication with them. This putative skill is said to be within everyone's capability (Targ and Puthoff 1977). For a period of time, this research seemed to promise a breakthrough in the search for a demonstrable psi effect. However, Marks and Kammann (1978, 1980), unable to replicate the remote-viewing effect themselves, discovered serious flaws in the remote-viewing procedure—flaws that they argued accounted for the observed effects.

The principal flaw concerned the judging procedure: Judges were asked to match up a series of responses against a set of targets. Marks and Kammann argue that because the transcripts of the subjects' reports were not edited to remove cues that would assist the judges in identifying the targets, the judging procedure itself—and not any psi effect—produced above-chance matching of transcripts with targets. Tart et al. (1980) responded to this criticism by first having the transcripts edited to remove any possible extraneous cues, and then having them rejudged. They reported that this did not eliminate the remote-viewing effect. However, Marks and Scott (1986), after obtaining access to the relevant materials (they had until recently been denied access to the raw data), report that the editing of the transcripts had failed to eliminate all the extraneous cues and that enough cues remained to account for the above-chance scoring rate.

There have been other criticisms of the remote-viewing studies as well, including concerns about statistical problems that could give rise to above-chance scoring rates (Hyman 1977b), and about the lack of adequate controls and control groups (Caulkins 1980). A number of replications and extensions have been reported (e.g., Bisaha and Dunne 1979; Dunne and Bisaha 1979; Schlitz and Gruber 1981; Schlitz and Haight 1984); only the Schlitz and Haight (1984) study appears to avoid the weaknesses of the Targ-Puthoff series, but even here, there was no control condition to allow proper assessment of the background "coincidental" scoring rate.

Thus, the Targ-Puthoff series is too flawed to be of evidential value, and none of the subsequent published studies have been carefully enough controlled to bear testimony about psi.

Schmidt's Random-event Generator (REG) Studies

For almost 20 years, Helmut Schmidt has been conducting research into the ability of subjects to predict or influence the radioactive emission of subatomic particles. His research enjoys generally high regard from other parapsychologists: Beloff (1980b), for example, views some of Schmidt's research as being among the most evidential in all of parapsychology, despite his own inability to replicate Schmidt's findings.

Schmidt has published a considerable number of studies. Unfortunately, this investigator typically completes a study and then— rather than focusing on a given research question, or refining his measurements, or examining the effects of various parameters in that particular situation, or working with one type of generator over a period of time so that he and others can come to appreciate its idiosyncrasies—he moves on to a totally different situation altogether (Hansel 1980), changing the design and components of his generator as he goes along (Hyman 1981). This makes it very difficult for him or anyone reading his research reports to learn the limitations of his generator or his procedures.

Little of Schmidt's research is free from serious methodological shortcomings (Hansel 1980; 1981; Hyman 1981). Consider, for example, one of his initial studies (Schmidt 1969a), which has been favorably cited many times in the parapsychological literature. The situation was as follows: A subject was seated before a panel of four lights and four corresponding buttons. On each trial, the subject would press one of the buttons to predict which light would next illuminate, something that would be determined by particle emission from a strontium-90 source. A light would then illuminate, giving immediate feedback. If the light corresponded to the depressed button, it was a "hit."

In the first experiment in this report, Schmidt combined the results from his three subjects and obtained a hit rate significantly higher than would be expected by chance: 0.261 as compared with 0.250 ($p < 2 \times 10^{-9}$). In the second experiment, subjects were allowed to choose to try to make a high or a low number of hits Here, the combined scoring rate of three subjects was 27 percent, again significantly higher than chance expectation ($p < 10^{-10}$).

Both experiments suffered from less than optimal experimental control; as in most of Schmidt's studies, subjects were usually unsupervised, and there was a general lack of rigor in the control of experimental conditions. Hansel (1980) objected to the fact that the exact numbers and types of trials undertaken by each subject were not specified in advance, and also to the fact that the equipment, although partially automated, did not rule out cheating during data classification.

There is a more fundamental concern about these experiments: the target series (Hyman 1981). Schmidt compared the subjects' hit rates to chance expectation, but this assumed that the target series was random. (Particle emission is presumably random; the output of his generator was not necessarily so.) Schmidt's randomization checks were carried out on target strings much longer than those used in the experiments, and therefore did not allow the detection of possible short-term biases in the generator that could give rise to nonrandom target strings. Because immediate feedback was provided throughout the experiment, and because subjects were free to "play" with the equipment and to decide when to start and stop a given session, any undetected short-term bias in the generator might give the subject the impression of being "hot" and therefore lead him to initiate a session, which he would probably end once he seemed to turn "cold." This, of course, could produce above-chance scoring rates.

It would therefore be important and appropriate to analyze the *actual* target sequence in terms of how well it conformed to what would be expected by chance. However, were one to find that the target sequence was nonrandom, this could, after the fact, be taken as evidence of PK. Indeed, Schmidt reported that after the testing one subject said he had tried to affect the outcome rather than just predict it; he had tried to produce more illuminations of lamp no. 4, he said. It was found for this subject that there was indeed an excess of 4s in his target series. No indication is given in the report as to whether this analysis of targets was carried out for other subjects, and if not, why not. However, Schmidt subsequently used this same piece of apparatus in a PK experiment (Schmidt and Pantas 1972) in which the only task was to try to influence the machine to produce an excess of 4s! Above-chance scoring rates were reported in that instance as well, which led Schmidt again to conclude that psi was operating. The skeptic is left wondering whether that apparatus simply produces an excess of 4s from time to time. Certainly, nothing can be concluded from such reports until more is known about the target series produced by the generator.

Thus, a study that seems at first to offer considerable evidence of an anomalous process is found to be badly flawed. It would make sense for Schmidt to redo the study, taking steps to make these criticisms unnecessary. Generally lacking in Schmidt's studies is a proper control condition: One should generate *pairs* of runs, with one run designated, on the basis of some random procedure such as the toss of a coin, as the experimental and the other as the control for each trial (Hansel 1981).

The problems in this study recur over and over in Schmidt's research (Hansel 1980; 1981; Hyman 1981). Only one of his studies appears well designed (Schmidt et al. 1986). However, we must wait to see whether the psi effect apparently obtained in this very recent study stands up to replication. There have been many psi studies (e.g., Targ and Puthoff 1974) in the past that at first appeared beyond reproach, only to be found later to be seriously flawed.

In summary, these various areas of research are plagued by methodological and statistical flaws of one sort or another. Until research is undertaken that is methodologically well planned and well executed—as Hyman and Honorton (1986) recommend with regard to the ganzfeld—there is little point in debating whether or not the existing evidence establishes a case for psi.

5
Does Parapsychology Follow the Rules of Science?

Of course, by using the term "rules of science," one could open up all manner of dispute because of the difficulty that exists in listing those rules and in demarcating science from pseudoscience (e.g., see Bunge 1984; Edge and Morris 1986). Rather than tackle that conundrum, it is more profitable to examine several aspects of parapsychological endeavor that appear to run counter to the spirit of scientific inquiry; each is discussed below.

Unfalsifiability

There are a number of principles in parapsychology that can be used to explain away failures to find empirical support for a hypothesis, thus creating a situation of unfalsifiability:

1. Perhaps the subject did significantly worse than expected by chance. If so, this may be taken as evidence of psi, because it seems to be *psi-missing*, something that occurs so often that it is now taken

to be a manifestation of psi (e.g., Crandall and Hite 1983).

2. If outstanding subjects subsequently lose their psi ability, or if subjects do more poorly toward the end of a session or of a series of trials, this is labeled the *decline effect* (e.g., see Beloff 1982). Rather than being taken as a possible consequence of either statistical regression or the tightening up of controls (when that has occurred), the decline effect often takes on the power of an explanation, because it has come to be viewed as a property of psi. For example, the decline effect in one experiment was interpreted as a "sign of psi" that was taken to strengthen the claim of a genuine psi effect (Bierman and Weiner 1980).

3. In a related vein, Schmeidler (1984) reports that PK effects are often strongest just *after* a session has terminated or during a subject's rest period. Rather than ignoring data accumulated after the session is over, this is taken to reflect another psi phenomenon, and has been given two names—the "linger effect" and the "release of effort effect." If this is to be taken seriously, then all researchers should report not only the presence of such an effect, but its absence as well: were this done, the frequency of the effect may well turn out to be within the bounds of normal statistical expectation.

4. Some parapsychologists seem consistently to obtain the results they desire whereas others are unable to find significant departures from chance (Palmer 1985b). The failure of one researcher to obtain significant results using the same procedure that yielded significant results for another researcher, rather than being taken as a failure to replicate or as a hint that extraneous variables may be producing artifactual results, is often interpreted in terms of the *experimenter effect*. This effect is so common in psi research that it has even been described by one parapsychologist as parapsychology's one and only finding (Parker 1978)! To *describe* the fact that two researchers obtained different results by calling it an experimenter effect is quite appropriate. After all, the experimenter effect as such is by no means unique to parapsychology, and a great deal has been written on the subject with regard to research in psychology and other domains (see Rosenthal and Rubin 1978). However, in psi research the term is

all too often used more as an *explanation* than as a description, and that is because it is considered that the effect may result not only from experimenter error (in that one experimenter may be more successful in obtaining psi effects than another because he unwittingly allows more artifacts to contaminate his procedure), or from differences in personalities (in that some experimenters may put their subjects into a more comfortable and psi-conductive frame of mind than others), but also from the psi influence of the experimenter himself (Krippner 1978a; Palmer 1985b; 1986b). If psi exists, of course, it would only make sense that the experimenter, who naturally wants his experiment to succeed, might unknowingly bring his psi influence to bear, whereas a skeptical or neutral experimenter might not use psi at all, or might use it to prevent the appearance of a subject psi effect. This whole problem leads Palmer (1985b) to describe the experimenter effect as the most important challenge facing parapsychology today. It is hard to imagine scientific inquiry of any sort if the results of the investigation are determined by the psychic influence of the investigator (Alcock 1985; see also Krippner 1978b).

The experimenter effect (or the experimenter psi version of it) provides a powerful method for undermining failures to replicate, and is sometimes resorted to for just that purpose. For example, when Blackmore (1985), a devoted parapsychologist for many years, found herself becoming increasingly skeptical about psi as a consequence of her inability to produce experimental evidence for it, she noted that "many parapsychologists suggested that the reason I didn't get results was quite simple—*me*. Perhaps I did not sufficiently believe in the possibility of psi" (p. 428).

In summary, it is the way such "effects" are used—and not, in principle, the research procedures—that vitiates the scientific respectability of parapsychology, for they make the psi hypothesis unfalsifiable by providing ways to explain away null results and nonreplications. These descriptive terms have mistakenly come to be taken as properties of psi, which leads to the circularity of explaining an observation by means of the label given to it. Moreover, as important propertiers of psi, their *nonappearance* in a psi experiment should weigh against any conclusion that psi has occurred; this never happens in the parapsychological literature.

All Things Are Possible

Another aspect of parapsychology that makes critics uncomfortable is what seems to be almost an "anything goes" attitude, with no speculation seeming too wild. For example, so-called observational theory based on paraphysical interpretations of quantum mechanics predicts that random events can be affected simply by being observed, even if the observation occurs at some time in the future (see Bierman and Weiner 1980). In line with this notion, studies have been done that claim to show that subjects can exercise an influence backward in time ("retroactive PK") so as to affect the choice of stimulus materials preselected for the study in which they are participating (e.g., Schmidt 1976). This also means, of course, that the present is possibly being influenced by future events (Martin 1983). A "checker effect" has also been postulated, in which ESP scores may be retroactively and psychokinetically influenced by the individual who checks or analyzes the data (Palmer 1978; Weiner and Zingrone 1986). Schmidt (1970c) reported that cockroaches were able to influence a random-event generator in such a way as to cause them to be shocked *more* often than would be expected by chance. He suggested that perhaps his own psi, fueled by his dislike of cockroaches, accounted for the increase, rather than a decrease, in shocks.

Not only can psi apparently transcend temporal boundaries; it also seems that no effort, no training, and no particular knowledge are required to use it. Indeed, modern PK studies appear to indicate that psi is an *unconscious* process, but a goal-oriented one in that it helps the individual attain desired objectives: Success in a PK experiment does not require knowing anything about the target, or even knowing that one is in a PK study (Stanford 1977). Thus, psi appears to operate very much like wishful thinking. For example, going back to the Schmidt (1969a) study, all that was needed, it seems, was for one subject to *wish* for a particular light to come on and it would light up statistically more frequently than the others. (Of course, when subjects do score above chance, neither they nor anyone else can say which hits were brought about by psi and which were the consequence of chance.)

As I have argued earlier (Alcock 1984), the fact that no physical variable has ever been shown to influence the scoring rate in psi experiments (Rush 1986c), combined with the apparent total lack of constraints on the conditions under which psi can be manifested (whether forward in time, backward in time, across thousands of miles, between humans and objects, between humans and animals, or even between animals and objects), serves to weaken the a priori likelihood that psi, as any source of force or ability, exists. After all, most psi experiments are very similar, in that all that is typically done is to examine two sets of numbers, representing targets and responses in an ESP experiment or outcomes and aims in a PK experiment, for evidence of a nonchance association. It may simply be that the enterprise of parapsychology generates, from time to time, significant statistical deviations—be they the result of artifact, selective reporting, or whatever—which are then independent of the research hypothesis, so that no matter what the researcher is examining—the effects of healing on fungus, PK with cockroaches, ESP across a continent, or retroactive psi effects—the likelihood of obtaining significant deviations remains the same. (For example, if an REG produces an excess of 4s on a short-term basis, and if the procedure allows subjects to tap into this, then it should make no difference in principle whether the targets are generated on-line or were recorded a week earlier: If the subject aims for more 4s, he will obtain them.) Difficulty in replication by other researchers using their own equipment or slightly different procedures would, of course, follow from such a state of affairs, as would the experimenter effect.

This psi-as-artifact notion is not offered as an empirically testable hypothesis. I only mean to show that the lack of constraints on the appearance of psi undermines rather than strengthens its credibility. It would be hard enough to accept that a philosopher's stone can turn base metals into gold, as alchemists believed. It would be harder still to believe that it can turn *anything* into gold and that anyone can use it without any training.

Lack of Rapport with Other Areas of Science

Parapsychology, despite its efforts to find common areas of interest with other research fields (see the *Handbook of Parapsychology* [Wolman 1977a]), has failed to establish any genuine overlap with other disciplines, because, so far at least, other disciplines do not seem to *need* psi. If "normal" explanations for strange physical or psychological phenomena were exhausted, and/or if the influence of the researchers' consciousness appeared to have an effect on the way matter behaved in "normal" experiments, then a much greater number of scientists might be more open to the possibility of psi. Indeed, if parapsychologists are right about psi, then the well-tested theories of physicists and neurologists are wrong (Hebb 1978). It is perhaps noteworthy that the claims that psi can influence radioactive decay do not come from particle physicists in the course of their everyday work.

6
Are the Critics Fair?

Some parapsychological proponents, such as Child (1985), argue that few in "normal" science bother to immerse themselves in the details of parapsychology, and instead gain a false or misleading impression from the accounts given by their colleagues who serve as critics of the field. Such critics are accused of unfair tactics, such as (*a*) arguing that unless fraud can be ruled out, it is the most parsimonious explanation of psi claims; (*b*) setting higher standards for parapsychological research than for research in the realm of normal science; and (*c*) simply rejecting the possibility of psi out-of-hand (see Collins and Pinch 1979).

Charles Tart (1982; 1984), a former president of the Parapsychological Association, suggests that there is an emotional basis for critics' unwillingness to welcome parapsychology into the scientific fold, an argument that has been repeated by Schmeidler (1985) and Irwin (1985b), among others. Tart posits that a widespread and unconscious *fear* of psi has developed either because strong psi ability would disrupt social functioning (because we would have access to one another's true feelings and thoughts) or because of what he calls "primal conflict repression": A mother often feels angry toward her child but keeps her cool and speaks to the child in a positive, supportive way. The child, if psi is already operating, is faced with a frightening conflict of messages and learns to represss psi altogether so as to avoid the information channel creating such conflict. Targ and Harary (1984), on the other hand, argue that skeptics base their opposition not on rationality but on religious conviction.

Suggestions about fear and religious conviction are too weak and ad hoc to require rebuttal. Collins and Pinch's (1979) concerns, on the other hand, are important. However, they could be equally relevant to any controversial claim, and thus nothing abnormal seems to be going on in the critical reactions to parapsychology. The scientific arena is a tough one; many ideas march in to do battle; some survive, but just as many perish. Numerous other controversial claims have faced hostility and even derision from scientists; some of these have won out (e.g., continental drift—see Hallam (1975); others (e.g., polywater—see Franks 1981) have not. Psychologists were at first unwilling to believe in the notion of biological preparedness with regard to learning (i.e., the idea that organisms, including humans, are biologically prepared to learn certain kinds of aversions more rapidly than others), and the leading journals refused to publish research reports on the subject, reports that are now viewed as being among the most important in their field (Seligman and Hager 1972). This concept is now part of mainstream psychology. Many psychologists also refused to believe in biological constraints on intelligence (or, at least, radically determined ones); and as a result of such dogged refusal to believe, the fundamental studies in this area—reported by Sir Cyril Burt—were eventually exposed as fraudulent (Kamin 1974). When, in the late 1960s, Neal Miller announced that he and L. Dicara had demonstrated operant conditioning of heart rate in rats (Miller and Dicara 1967), many experimental psychologists refused to believe it, despite Miller's good reputation as an experimental psychologist. Ultimately, Miller himself, when subsequently unable to replicate his own studies, publicly withdrew his claims (e.g., Miller 1978). *Science* refused to publish, on the grounds that it was erroneous, the initial research of Solomon Berson and Rosalyn Yalow (Yalow subsequently won the Nobel Prize) on the insulin-binding antibody, research that was fundamental to the development of the radioimmunoassay technique (Garfield 1986; Yalow 1978). Albert Einstein absolutely refused to believe that "God plays dice," despite the implications of quantum mechanics; he chose to believe the theory to be in error due to incompleteness. Science is full of such examples.

This process, although sometimes seemingly cruel and dogmatic, is perhaps necessary to allow scientists to focus on claims that appear most promising, rather than being distracted by others that appear to have little to recommend them. Sooner or later in science, it seems, the truth will out, and error falls by the wayside. Even acupuncture, long regarded as being nothing short of superstition, is now regarded as capable of producing limited pain relief (Zusne and Jones 1982).

If the insulin-binding antibody, biological preparedness, and acupuncture analgesia won accommodation in science, it is because the evidence for them became so strong that they *had* to be accommodated. A century of parapsychological research has gone by, and the evidence for psi is no more convincing than it was a century ago.

It seems accordingly that parapsychologists who attack scientists and critics for their refusal to recognize the importance of psi and of psi research are attacking the messengers because they cannot accept the messages they bear. Suppose that instead of psi, parapsychologists were promoting a cure for baldness, but that the amount of hair produced by the treatment was tiny and detectable only by some researchers, sometimes. If the effect is unreliable and unrepeatable, if it also contradicts all that is known about hair growth and alopecia, and if there is no theoretical mechanism put forth for the putative effect, then one would hardly expect the scientific community to cheer the end of baldness. Science will never take parapsychologists simply at their word; they must offer a clear, replicable demonstration of a basic phenomenon in order to gain acceptance in science.

Moreover, one can seriously challenge the claim that practitioners of normal science do not give, or have not given, parapsychology its day in court. As was mentioned at the outset, a number of professional scientific organizations have invited parapsychologists to address them or have set up symposia on the subject. True, parapsychological ideas have hardly been embraced with open arms, but that does not mean that scientists are motivated by fear or blind prejudice or ignorance or distorted interpretations purveyed by unreliable skeptics.

Indeed, when parapsychology began to take shape as a serious research field, a good number of psychologists and others immediately took up the challenge of investigating claims of spiritualistic communication, telepathy, clairvoyance, and so on. All that was lacking to make parapsychology part of mainstream psychology was evidence that there was a phenomenon to investigate. At the Fourth International Congress of Psychology, held in Paris in 1900, an entire section was devoted to psychical research and spiritualism, and the president, T. A. Ribot, announced the founding of a psychical research institute in Paris (L'Institut Général Psychique) (McGuire 1984). Membership in this institute included a number of prominent psychologists, such as Janet, Richet, James, and Tarde. In 1895 Binet published some case studies of telepathy. However, as McGuire (1984) points out, psychologists were already becoming very uneasy about the growing link between psychical research and spiritualism; this mistrust began to show itself at the Fourth Congress, and subsequently many French psychologists began to turn their backs on psychical research.

Psychologists Pieron, Janet, and Dumas conducted a number of séances in which they reexamined mediums who had produced positive outcomes in earlier studies at the Institut Métapsychique. One medium was caught flagrantly cheating, and these psychologists concluded that no psychical phenomena had been observed under the carefully controlled conditions. LeBon offered a large reward to anyone who could produce the mediumistic effects in his laboratory, but once informed of the stringent controls, no one ever underwent the test (McGuire 1984).

The American Society for Psychical Research was set up in 1885 to examine apparent psychical phenomena (Moore 1977). Its officers included prominent psychologists, such as Prince, Hall, Jastrow (later to become an outspoken critic), and James. When they failed to find any evidential basis for mediumistic claims, most members lost interest; the group was disbanded, and its remnants merged with the British Society for Psychical Research (SPR) in 1889. James continued to support and believe in psychical research, and later became president of the SPR. (The American Society for Psychical Research was reestablished in 1907, but it no longer had the critical and tough-minded attitude that characterized the original organization.)

In the 1930s, parapsychology had another opportunity to persuade mainstream science about the importance of psi research. A poll conducted in 1938 found that 89 percent of psychologists felt the study of ESP was a legitimate scientific enterprise and 79 percent felt such research was a proper subject for psychologists (Moore 1977). In that same year, a round-table discussion of parapsychology was sponsored by the American Psychological Association. Parapsychologists did not succeed in their attempts to gain the psychologists' support for the study of psi.

The 1970s provided another period when mainstream science seemed ready to give parapsychology a chance. As mentioned earlier, the Parapsychological Association had gained affiliation with the American Association for the Advancement of Science in 1969. In 1974 one of the world's leading scientific journals, *Nature*, published an article by paraphysicsts Targ and Puthoff in which they detailed their claims about scientific evidence for the paranormal, based largely on research with Uri Geller (Targ and Puthoff 1974); true, the journal did precede the article with an editorial disclaimer, but the reasearch nonetheless appeared. Although some parapsychologists were irked by the editorial "inoculation" *Nature* provided for its readers, such a disclaimer proved to have been prudent, because, as discussed earlier, Uri Geller was subsequently exposed as a trickster (e.g., Randi 1975).

Although mainstream psychological journals continue to be reluctant to publish parapsychological research, that is not to say that these journals are totally closed to parapsychologists; occasionally articles do appear (e.g., Layton and Turnbull 1975). *American Psychologist* recently published an article (Child 1985) that presented, along with his criticisms of skeptics' interpretations of parapsychological research, the results of a meta-analysis of the classic Maimonides dream studies. Child concluded that something important is going on, although, in my view, his analysis is unlikely to impress many psychologists. Parapsychology was discussed in an open-minded fashion, albeit very briefly, in a recent issue of the *Annual Review of Psychology* (Tyler 1981). Since 1950, more than 1,500 parapsychological papers have been abstracted in *Psychological Abstracts*, which is published by the American Psychological Association (McConnell 1983).

What more should parapsychologists expect, given the track record they have produced? I am of the strong opinion that rejection of, or dissatisfaction with, paranormal claims is not based on narrow, dogmatic prejudice, but on the fact that after a century of research, there is still nothing substantive to show!

7
Is Rapprochement Between Psychology and Parapsychology Possible?

In 1982 psychologists Zusne and Jones's *Anomalistic Psychology* was published. This book constituted a milestone in the course of interaction between psychology and parapsychology by virtue of its attempts to establish a framework for the psychological study of the phenomena taken by parapsychologists to be paranormal. Blackmore (1983a), coming from the parapsychological side, and just as she was renouncing her belief in the psi hypothesis, also called for the study—within psychology—of the experiences that appear to people to be paranormal. Palmer (1986a) calls for a collective focus by skeptics and parapsychologists on finding explanations for anomalous experiences and phenomena, whether the explanations prove to be mundane or not. These actions may reflect what Truzzi (1985) views as a movement toward rapprochement between psychology and parapsychology.

Unfortunately, I doubt that such a rapprochement will ever occur, for I believe that those in parapsychology who move closer to the skeptical side will fail to draw the rest of parapsychology along with them. That is not to say that there will not be cooperation between psychologists and parapsychologists in the study of anomalistic experiences, something that should be strongly encouraged; nor is it to deny whatever movement there has been toward better mutual understanding and respect.

However, finding explanations for ostensible anomalies is not what parapsychology is really about for most parapsychologists. If it were, much more effort would be made to try to find psychological and neuropsychological explanations for such experiences before even contemplating the radical psi hypothesis. (Indeed, one must wonder why parapsychologists seem not to concern themselves with the actual *experience*, or with *how* such experiences are generated, or with *how* the supposed phenomena work [Scott 1985]. Why, for example, do they not set out to try to produce in subjects the subjective impression of telepathy, instead of merely concluding that subjects in a guessing task must have experienced telepathy on some of the trials? Studying guess rates is *not* the study of the telepathic *experience*.)

If parapsychology is not primarily motivated to explore anomalies in an open-minded fashion, what *is* its motivation? Why does parapsychology persist after a century of failing to produce compelling evidence of psi? Why does the psi hypothesis survive? To be fair, of course, normal science does not reject working hypotheses just because they fail to be confirmed empirically—although they rarely, if ever, show such longevity. For example, Harvey's theory of the circulation of the blood depended upon the existence of capillaries, and such capillaries could not be observed with the naked eye; but the failure to observe them did not lead to rejection of the theory. Investigators continued to seek them until, with the aid of microscopes, they were at last discovered (Gregory 1981). Yet, there is a difference between, on the one hand, not giving up a preferred hypothesis when that hypothesis seems to promise more explanatory power than existing theories about a range of observations and, on the other hand, the discounting of failure to find expected statistical deviations in a psi experiment. In the latter case, one is trying to establish the existence of a phenomenon that is *not* required by the existing body of scientific data, nor is it predicted by theory, nor would it simplify or clear up current anomalies in physics or psychology or biology.

The dispute about psi reflects the clash of two fundamentally different views of reality. The first of these is the materialistic, monistic view that the human mind is some sort of emergent manifestation of brain processes, whereas the second is the dualistic position that maintains that the human mind/personality is something beyond the stuff of atoms and molecules. Parapsychology grew out of the second of these; it developed directly from attempts, both in Europe and the United States, to put the postmortem survival of the human personality on a sound scientific footing (Cerullo 1982; Mauskopf and McVaugh 1980; Moore 1977). It is the search for the soul— not the Soul as it is described by various religions, and perhaps not even the secularized soul sought by the psychical researchers of the late nineteenth century during the heyday of spiritualism (Cerullo 1982), but a soul all the same. Because, if the mind can operate separately from the physical brain, as the psi hypothesis would suggest, then it possesses much of what has been ascribed to the soul.

Most religions teach that the Soul survives death in some form. The questions of survival of the parapsychologists' "soul" or "mind" or "personality" after death is, even many leading parapsychologists agree, an important question for parapsychology to consider (e.g., Krippner 1983; Palmer 1983; Roll 1982). Blackmore (1983b) suspects that just as it was the *fundamental* question to many of the early psychical researchers, it is still so today.

Thus it is important in any debate about parapsychology to make clear just what is being debated. Is the debate about whether or not there exist "natural" phenomena that science has so far failed to recognize, or is the debate about whether or not dualism, as opposed to materialistic monism, is the correct view of nature and of mankind's place in nature? Or, is the first question very often the surface issue, while the hidden agenda is the question of dualism?

Conclusion

"Either parapsychology is a harvest of false illusion, or the meat and fiber of biology, the focus of psychology, and even the material conception of physics on which all science stands" (Walker 1984, p. 9). These words by a paraphysicist should remind us that the existence of psi is no trivial matter. Yet, to accept the reality of psi, we must accept that some force or process exists that cannot at this time be described in terms of positive properties, but only in terms of what it is not; a force that is capable of allowing for direct communication between two brains, regardless of the distance between them, and that allows the mind directly and often unconsciously to influence matter in such a way as to gain some desired goal, again without any effect of distance, physical barriers, or even time. To accept the reality of psi, we must discount a hundred years of failure to find substantive evidence; there is not a single demonstration that is repeatable in Beloff's "strong sense." We must also accept that there are fundamental problems with well-tested physical and neurophysiological theories. We must accept all this in the face of the inability of parapsychologists to sort out whether, in a given experiment, a statistical deviation is due to PK or to ESP, whether it is due to the subject or to the experimenter, and whether the source of psi is acting in the past, the present, or the future. Futhermore, we must overlook the fact that even the best research programs in parapsychology are seriously beset by methodological weakness. We must ignore history as well, for as Hyman (1981) points out, each generation of parapsychologists has put forth its current candidates as providers of proof of psi— experiments that supposedly should have convinced any rational person were he to examine the evidence fairly. Yet, these candidates keep changing, and if prior history is a reliable guide, today's most promising research programs in parapsychology may well be passé in a generation or two.

If parapsychologists really are dedicated to the study of anomalous experience, then it should make more sense to follow Blackmore's (1983a) lead and focus on the anomalies while putting the concept of psi aside until, if ever, it is needed. This is unlikely to happen, however. Psi has been postulated not because normal psychology is incapable of accounting for people's apparently psychic experiences, nor because of inexplicable findings in physics or chemistry; nor is it the logical outgrowth of some compelling scientific theory. Rather, the search for psi is now, as it has been since the formal beginning of empirical parapsychology over a century ago, the quest to establish the reality of a nonmaterial aspect of human existence—some form of secularized soul.

All that is needed to turn the attitude of the scientific establishment from doubt to serious interest with regard to psi is to produce some clear, substantive evidence of a psychic phenomenon. Without it, parapsychology can never become a science.

Postscript: A To-do About Dualism Or a Duel About Data?

The following is my reply to the many commentaries on the foregoing article that were published in the same issue of Behavioral and Brain Sciences *(vol. 10, no. 4, December 1987).*[1]

*   *   *

It is a humbling experience to be faced with nearly 50 commentaries from some of the world's most respected advocates and critics of parapsychology. Their responses themselves reflect to a degree the controversy that surrounds parapsychology: My target article was at one and the same time adjudged to be "brilliant" (Mario Bunge), "excellent" (Henri Broch), and (along with that of Rao and Palmer [R and P]) "outstanding" (Stanley Krippner), and to be "misleading and inaccurate" (Charles T. Tart), hyperbolic (Peter Railton), and ad hominem (Marcello Truzzi). I shall try to deal with the criticism; let others take issue with the praise! In so doing, I have organized my response around a number of themes that recur in the commentaries.

Worldview

Among those who reacted negatively to my paper, the most common criticism, made by about one-quarter of the commentators (including some who otherwise share my skepticism), was with regard to the subject of dualism. I am described as being "obsessed with dualism" to the extent that it is difficult to evaluate my arguments about the evidence for psi (John T. Sanders), as focusing too much on the motives of parapsychologists, (Victor A. Benassi), as considering parapsychology to be a thinly disguised search for a metaphysical ideal and not really a science at all (Charles Akers, Richard S. Broughton, Brian MacKenzie, Charles Tart), and as advocating the materialism of the dominant (orthodox) psychological outlook through an essentially ad hominem attack on parapsychologists (Truzzi). Nicholas P. Spanos and Hans de Groot suggest that from the title onward my article assumes that parapsychologists' supposed dissatisfaction with materialism *ipso facto* makes their endeavors scientifically suspect, a view shared by Tart.

Yet very little of my target article actually had anything to do with this subject: Apart from the title and abstract, and a comment about how the *concept* of paranormality implicitly involves mind–body dualism, there are only two or three paragraphs at the end of the article that focus on the subject of dualism, and these relate not so much to the mindset of parapsychologists as to my speculation about the *persistence* of parapsychology and to my views on the nature of the dispute about psi.

There was clearly some misunderstanding (and obviously I must take the blame for that) about just what it was that I was saying when I discussed a search for the soul. Sanders, for example, took my arguments to be directed at the claim that otherworldly souls have been demonstrated; because Rao and Palmer do not make such a claim, he said, my arguments do not strike home. However, I did *not* suggest that any such claim had been made, nor do I believe that the large majority of parapsychologists are trying to demonstrate the existence of disembodied "souls" as such. My contention is that there is an implicit search for something that lies outside the scientific worldview as we know it today and that that "something" involves an ability of the mind to interact directly with other minds and with matter, and that therefore implies an aspect of human personality that is not tied to the body. I used the term "search for the soul" as a metaphor for this.

I similarly used the term "mind–body" dualism as a label for the notion that the mind can act independently in some manner, whether leaving the body, as some suggest, or interacting directly with nature, and this provoked criticism from several commentators. For example, Stephen E. Braude accuses me of failing to grasp important issues about reductionism and dualism and never acknowledging that cognitive or intentional phenomena generally—normal *and* paranormal—might simply lie outside the domain of the physical sciences. David Navon points out that the philosophical doctrine of mind–body dualism does not encompass Rao and Palmer's concept of omega [a term they use to identify potential explanations of psi that go beyond the generally accepted scientific view of what is possible in this world], because mind–body dualism could imply an autonomous psyche, a psyche that is affected by or interacts with the body, but not the idea that a person's mind or spirit can interact with the environment while bypassing the body.

I regret having used the terms "soul" and "mind–body dualism," for they obviously failed to communicate what I was trying to say. Moreover, there is plainly much more to the metaphysics of dualism than my simplistic use of the term suggested. Perhaps if I had used a term such as "a radical dualist (interactionist) position" (as John Beloff did), or if I had simply stuck to a description of the contrast between parapsychology and "normal" science as a competition between worldviews (Jerome J. Tobacyk) or metaparadigms (Navon), I would have avoided difficulty. It is unfortunate if the dualism issue distracted some commentators' attention away from my essential arguments, which concerned methodology and the interpretation of data.

However, I reiterate my view that parapsychology reflects a worldview that opposes the predominant materialistic worldview of contemporary science. I am supported in the view by no less an authority than Beloff when he states in his commentary that psi contradicts physicalism, the doctrine that everything must ultimately be explainable in terms of physical laws. Consider Helmut Schmidt, for example. Although he rejects the claim that his work is an attempt to establish the reality of a nonmaterial aspect of human existence, he then admits that he is attempting to show that current physics fails in systems that involve human subjects. This comment would support my (unfortunate) "soul" metaphor, because if mind/thought/personality can interact with matter independently of the body, it would seem that the personality in some way exists separately from the body.

Some commentators read into my remarks that I believe parapsychology to be unscientific on the basis of this presumed difference in worldviews. Benassi and D. C. Donderi interpret me as suggesting that parapsychologists fall prey to tricksters and conduct experiments because they are not searching for scientifically validated truth but for the soul. Tart argues that interest in the notion of a soul is not inherently unscientific in any case, and William R. Woodward argues that monism and dualism are metaphysical positions allowable to one making any scientific claim. Yet nowhere did I say—nor would I suggest—that parapsychologists are poor scientists simply because they take a dualistic or any other metaphysical position. Indeed, it seems to me that some commentators were so upset by any reference to "soul' or "dualism" that they took for granted that I was being pejorative. If it is inherently "unscientific" to be searching for evidence of either, let them say that themselves.

Donderi argues that I postulate a priori a materialistic universe that precludes the existence of paranormal phenomena, and Tart feels that it is my certainty about, and attachment to, current physical theories that stimulates my attack on parapsychology in the first place. Yet nowhere did I state that the dualistic hypothesis has to be wrong or that it is to be eschewed by those who practice science. Nor am I emotionally unable to accept some kind of dualism, should evidence and logic demand it. Either reality is dualistic or it is not. Either psi, in R and P's omegic sense, exists or it does not. Ironically, Irvin L. Child sees me as the theory truster and R and P as the observation trusters. I happen to believe that exactly the opposite is the case: I cannot find any data that are persuasive, and therefore I see no reason to change my materialistic worldview or to believe that psi exists. I would argue that if R and P *are* data-driven, then, unless there are some substantial breakthroughs in the near future, we should expect them to follow in the footsteps of other parapsychologists, such as Susan Blackmore and Charles Akers, who have in a very real sense moved outside parapsycholgy to become critics.

I want to stress that I did not by any means intend to imply that parapsychologists want anything less than to find "scientifically validated truth." Of course they do; the argument is whether or not they are following sound scientific procedure in their attempts to do so—whether, for example, because of unfalsifiability of hypotheses (through, say, the psi version of the experimenter effect), they have gone astray. Nor should any parapsychologists' data be disregarded because of any particular worldview. In the past I have written (Alcock 1985, p. 561) that

> it is important to note that it is not because of Newton's testimony or Newton's beliefs that his classical laws are accepted as principles underlying certain aspects of nature on a macro level. It is not even because of his data, some of which was fudged (Westfall 1973). Nor is it because of Einstein's philosophy that most scientists now believe that his theory of relativity is substantially correct. It is not because of Pasteur's beliefs or motivations or even his data that we believe that pasteurization of milk is efficacious and essential. Clearly, personal motivations are irrelevant here; in all these cases, we believe because subsequent empirical testing of these ideas has repeatedly supported them. On the other hand, we would be wise to refuse to accept any of them if only people who were already persuaded of their reality could find empirical support.
>
> In parapsychology, as in these cases just cited, the motivations of the researchers are also irrelevant to the evelution of their claims. They become important only in our understanding of the *persistence* of the quest. (Italics in original.)

I continue to hold that view, and I am sorry if that did not come through clearly in my target article. On the other hand, Gary Bauslaugh is right in stressing the importance of credibility in those cases where *only* believers in a phenomenon are able to produce evidence for it. If the only scientists whose data go against the view that smoking causes cancer are those who work for tobacco companies, then we may have reason for concern. If the only researchers who are able to detect psi are those who are persuaded a priori that psi exists, then perhaps we also have reason for concern. All may be excellent scientists; all may inadvertently bias their data and their interpretations as a result of their expectations. The same, of course, applies to those, including skeptics, who may have other belief orientations. That is why the crucible of scientific debate is so important and why replicability by others who do not share the a priori belief assumes such a prominent role in such cases.

Before leaving this subject, I believe that the relationship of belief systems to parapsychological endeavor is one that is worthy of empirical examination. Yet, as Krippner correctly points out, I have no hard data to support my speculation about the reasons for the persistence of parapsychology. In this vein, Akers refers to Allison's (1973) survey (which was unknown to me because it has not been published) of the membership of the Parapsychological Association in which 43 percent disagreed with the statement that "the results of parapsychological research clearly indicate that there is a nonmaterial basis of life or thought." He suggests that those data undermine my argument. I would disagree. First, Allison's data refer to the evaluation of the evidence, not to the motivation to pursue parapsychology. Second, I would imagine that someone like Beloff—who in his commentary argues that the importance of parapsychology is that it alone can provide the relevant empirical evidence in deciding between an epiphenomenalist and a radical dualist position on the mind-brain relationship—would disagree that the evidence *clearly* supports that view (see his commentary). Therefore, those who disagreed may well, as Akers did acknowledge, simply be unconvinced by the evidence. What is more important, I would argue, is that 56 percent of the Parapsychological Association membership actually *agreed* that the evidence is already in. To me that is astounding. Over half the membership in the leading professional parapsychology body is already persuaded about a nonmaterial basis for life or thought.

The Problem of Definition

Several commentators took up the issue of the definition of psi. Indeed, one difficulty that I have encountered in preparing this response is that most commentators seem to implicitly implicate the paranormal when they refer to psi, which is quite out of keeping with the approach that R and P formally advocate, albeit in line with what most parapsychologists themselves do in their own writing. Although psi was originally conceived as a neutral term for paranormal phenomena, it now seems that one must differentiate between "psi" and "parapsi."

Sanders argues that R and P have tried to observe the convention, coming, he says, from Ray Hyman and Charles Honorton (1986), that psi is to be understood as a communication anomaly. He contends that I have failed to do the same and that it is thus difficult for him to see how my arguments about dualism can be brought to bear. However, psi as a communications anomaly has yet to become the conventional usage. Hyman points to this when he refers to R and P's definition of psi in terms of anomalies as "unorthodox"; R and P went on to qualify this definition by referring to anomalies as "ostensibly paranormal." Indeed, as both Railton and Akers observe, paranormality crept back into R and P's concept of psi even after they had formally thrown it out. Railton shows how the concept of psi used both by Hyman and Honorton (1986) and by R and P departs from simple anomaly and leads into the paranormal.

R and P introduce a new term, "omega," to which they assign the paranormal aspects of psi. However, as Navon remarks, this new term is unlikely to be helpful because it is another blanket label that can be applied to a wide range of phenomena. I agree with Rex G. Stanford's suggestion that this term not be used, for it is already loaded with other meaning that will only serve to further confuse. There is more than a little irony in the fact that, whereas I discussed the persistence of parapsychology in terms of a search for the soul, R and P chose the term "omega," with its religious and spiritual connotations usually associated with death (as in *Omega: The Journal of Death and Dying*), to describe the paranormal.

How can the 56 percent of the membership of the Parapsychological Association who believe that there is already clear evidence for a nonmaterial basis of life or thought be content with a definition of psi that directly undermines that view by speaking only of anomalies? Adrian Parker makes the point clearly: By redefining the field as the study of ostensible psychological anomalies, one may blur and compromise the issues. He seems to be saying that this would be akin to throwing the baby out with the bath water, for he believes that the past 50 years of research has done more than simply point to anomalies of one sort or another. Parker wonders whether parapsychologists would not do better simply to define parapsychology as the study of phenomena that apparently relate to the nature of the interaction between consciousness and the brain by suggesting that consciousnes can directly influence external events. This would certainly make it much clearer just what is being discussed.

Anomalies and Explanation

Obviously, the crucial question is whether or not there are "communication anomalies" that require explanation. In Child's opinion, R and P have made a clear case for the presence of anomalies. In Hyman's opinion, on the other hand, this case remains to be established, and he sensibly calls for such anomalies to be referred to as "ostensible" psi anomalies, a term that by coincidence Palmer introduces in his commentary. I want to underscore the fact that there have, to my knowledge, never been anomalistic claims, from either mainstream psychology or natural science, that seem to imply psi or the paranormal. (Incidentally, I find strange Trevor Pinch's recommendation that in order to win recognition from mainstream science, parapsychologists should demonstrate that equipment from other areas of science does not work as expected or needs to be modified, or they should convince orthodox psychologists that they should control for psi effects in their own research. Surely this is begging the question.)

Some parapsychologists argue that psi, whatever that means, is at least as believable as the explanations that critics contrive, if not more so. Parker, for instance, believes that "the probability for *all* these contrived explanations being valid is beyond the limit of (my) common sense." Palmer expects skeptics to provide specific explanations for psi data, and argues that conventional alternatives are so deficient that even their strongest proponents are reluctant to acknowledge them as interpretations of the data. I do not know what he means by this, for as Beloff remarks, skeptics do not need to demolish R and P's evidence, nor do they need to provide plausible counterexplanations, nor are accusations of fraud necessary. All that is necessary is to point to the methodological shortcomings of the experiments. In the same vein, Hyman argues that there is no need to try to provide conventional explanations, because parapsychologists have yet to demonstrate the existence of anomalies. Hyman makes the excellent further point that in the absence of adequate theory and reasonable specifications, there is no reason to expect that an alleged departure from chance baseline in one study is the same "thing" that produced another alleged departure from chance in another experimental situation.

Akers shares my view that psi effects arise from the normal "garden variety" artifacts that appear in unguarded moments in other social-science research. Truzzi, on the other hand, thinks that something "interesting and probably new" is going on, but suggests that this may be due to new kinds of artifacts yet to be discovered. It is not clear to me how Truzzi would find reason to go beyond Akers's position at this point.

Questions of Theory

Two major underlying concerns regarding theory and parapsychology emerge from the commentaries. On the one hand, there are those who argue, as does Bunge, that genuine sciences are members of a closely knit system of partially overlapping research fields, and that new knowledge claims should be viewed with some suspicion if they are inconsistent with what is already taken to be known. Bunge, Navon, and others argue that the omega hypothesis conflicts not just with the data of science but with its foundations. On the other hand, there are those who, like Victor Adamenko and Jessica Utts, worry that a focus on such consistency will lead us away from important discoveries that could radically change that worldview. However, the notion of psi had been around long before many of the "impossible" things of the past became "possible," and it rests much as it did a century ago on very controversial evidence. Both psi waves and gravity waves may someday become part of scientific "reality," but as Bunge eloquently points out, the latter have the advantage of already being consistent with a considerable body of theory and data. Despite O. Costa de Beauregard's radical contention that the theoretical formalism of modern physics implies paranormal phenomena, a view that Brian D. Josephson seems to support, I believe that most scientists would agree that the fact that psi does not "fit in" with existing knowledge, or with our beliefs about causality, is good reason for assigning it a low a priori likelihood of existence. While we are on the subject, Braude argues that I am completely wrong in suggesting that it is a logical principle that a cause cannot precede its effect. However, another philosopher, Antony Flew, argues that backward causation admits to conceptual incoherence and self-contradiction.

Barry Beyerstein provides an excellent review of some of the difficulties that psi (of the paranormal variety) poses for neuropsychology, or, more to the point, that neuropsychology poses for psi. I urge parapsychologists to take his comments very seriously. The body of evidence built up over the years by psychologists and neurologists about the functioning of the mind/brain cannot simply be bypassed or dismissed. Beyerstein points out that the monist position was at one time a rather radical one that came gradually to be accepted because of the weight of the evidence, and not because researchers wanted to avoid dualism. Beyerstein's commentary also, I believe, speaks to Parker's suggestion that parapsychology is clearly in tune with the contemporary view in psychology that consciousness and experiencing have a steering function in the organism. The contemporary view gives no succor to parapsychological hypotheses.

Questions of Methodology

Each research discipline develops its own methodology or borrows from that of its neighbors. Often, the methods that evolve in one discipline are not applicable in other disciplnes. For example, the controlled laboratory experiment as used by psychologists is totally inappropriate to the study of astronomy; the methodology of chemists is of little relevance to a field botanist. This gives rise to two very important considerations:

1. We cannot judge whether a field of research is "scientific" simply on the basis of its methodological approach. There must be an appropriate match between its methodology and its subject matter (Hyman, personal communication). Tart suggests that there is nothing unscientific about a search for the soul. I believe that although we should not, of course, decide a priori that the question is not scientific, this claim cannot even be evaluated without looking at the methodology involved. Using radio telescopes to scan the heavens for the souls of our ancestors would presumably not be scientific, unless there was some good theoretical reason to think that souls might be located by such a method. Presumably, then, one should begin with some axioms about the nature of the soul, formulate hpotheses on this basis, and then test the hypotheses using techniques that appear appropriate given what one believes about the phenomenon. What axioms do we have in the case of psi? What methodology would be appropriate? If psi involves the paranormal, then it may be that the methodology of experimental psychology that has been taken over by parapsychology is inappropriate. For example, Robyn M. Dawes calls for experimental control of the conditions that are supposed to produce or enhance psi. This is what one would try to do in mainstream psychology, but Dawes recognizes the virtual impossibility of doing this with psi because there is no body of theory

that specifies when psi should not occur. Braude recognizes the problem as well when he argues that the conventional experimental methods in parapsychology are powerless to reveal anything interesting about the phenomena, except perhaps that they exist, because if we take psi (i.e., the paranormal) seriously enough to test for it, then we must give up the ideal of the blind or double-blind protocol. I would go further than Braude and argue that to demonstrate even the very existence of paranormal phenomena is problematic for the same reason.

2. Because methodology tends to be discipline specific, and because the sources of possible error are somewhat different from discipline to discipline, it is not always easy for someone trained outside the discipline to evaluate the methodology used within the discipline. Geologists are not concerned with placebo effects; therefore, if a geologist were to evaluate clinical studies of a particular drug, he might take the study to be carefully designed and executed even in the absence of controls for placebo effects. Physicists and philosophers have little experience in working with human subjects and can quite often be very naive about the many ways in which subtle influences can confound data (just as some psychologists can appear naive about philosophical concepts like mind-body dualism!). Moreover, because effect sizes are typically so small, and because even minimal communication between subjects in ESP experiments must be eliminated, there are special problems of experimental purity in parapsychology that many psychologists know little about. My point here is that scientists and philosophers should not automatically assume that they can competently evaluate the findings of parapsychological research unless they have done the necessary homework and carefully explored the methodological traps and statistical pitfalls specific to parapsychology. It is not enough to let the data speak for themsleves; they always need interpretation. Thus, when Broughton argues that many scientists—including quite a few whose credentials in more orthodox fields are unassailable—read the same parapsychological reports I read and come to the conclusion that there is something to be investigated, I find that general comment no more informative than if he were to say that outstanding geologists and astronomers

have passed judgment on medical controversies. That is not to denigrate these scientists; it is simply to say that expertise in one area does not automatically lead to critical acumen in another.

With regard to alleged flaws, Broughton chides me by saying that all experiments everywhere are flawed. That may be true. However, as Blackmore points out, even if parapsychology's methods are often as good as psychology's (which, incidentally, may be because of weaknesses in psychological research), the difference is that psychology can tolerate considerable error and still progress. Remember, too, that the essential claim of the parapsychologist is that *all conventional sources of influence that might have brought about a higher than expected correlation have been ruled out*. If they have not, then by definition there is no anomaly. Conventional psychological research certainly does not involve the demonstration of phenomena that otherwise seem to conflict with the scientific corpus of knowledge and that in order to be demonstrated require such stringent control of extraneous variables, not all of them necessarily known at any given time.

More and more, parapsychologists are entering into debate about whether or not a flaw is serious enough to vitiate the results of an experiment. Sometimes it is argued (as by Palmer) that the onus should be on those who point to the flaw to show empirically that it could in itself produce the effect, and that flaws found by critics need to be evaluated against the same standards of plausibility and empirical evidence as any other scientific explanation. In other words, the critic should be able to demonstrate that the flaw could have produced the outcome. I can think of no area other than parapsychology where anyone has attempted to place the onus on the critic to demonstrate that an acknowledged flaw was both a necessary and sufficient cause of the effect. Psychological researchers do not argue over whether or not a particular flaw was basis enough for the results. I presume that physicists and chemists and medical researchers do not argue, particularly when controversial claims are involved, that the flaws were not bad enough to produce the observed effects. Presumably, they repeat the study *without the flaws*. Why should anything less be expected of parapsychologists?

The Schmidt Studies

I am accused by Schmidt of never having read his paper firsthand, a point that is amplified by Broughton, who accuses me of depending on Hyman and Hansel. In reality, I have carefully read all the published papers of Schmidt of which I am aware. However, I do apologize to Schmidt for my remarks about lack of supervision. Although there was inadequate supervision of subjects in some of his studies (Schmidt 1970c; Schmidt 1978; Schmidt and Pantas 1972), this has not been a general problem, and I was in error in suggesting that it was.

With regard to the source of randomness, Schmidt indicates that he has used the same type of generator for nearly all his experiments, although he admits that there have been changes in the circuitry as time has gone on. In fact, Schmidt (1969a) used a modulus-4 random-event generator (REG) based on radioactive decay. Schmidt (1969b) used the Rand tables. Schmidt (1970a; 1970c; 1978; 1979a) used a binary output REG based on radioactive decay. Schmidt and Pantas (1972) went back to the modulus-4 REG/radioactive decay. Schmidt (1973; 1976) used an electronic noise REG. Schmidt (1974) compared two different REGs. Schmidt (1979b) used an electronic die based on radioactive decay. Schmidt (1981) used seed numbers for an algorithm that were generated by a binary REG based on radioactive decay. Schmidt (1985) and Schmidt et al. (1986) used a computer that was driven by radioactive decay. So, although Schmidt has been pretty consistent in using radioactive decay (or sometimes electronic noise) as the ultimate source of randomness, the apparatus used to "capture" this randomness has varied considerably, leading me to reiterate that the REG, which is *more* than just radioactive decay, has varied from experiment to experiment. Not only is a given generator not explored in a consistent way, but the series of experiments themselves do not build upon one another. Furthermore, in many of these studies, there are clear methodological shortcomings apart from concerns about randomness. For example, Schmidt has often served as his own subject, and in one case was really the only subject. In many of the studies, there are varying numbers of trials or sessions per subject, and these are usually combined. Schmidt has generally worked in isolation from others, and his raw data, with little exception, have not been available for scrutiny. Martin Gardner urges that Hyman be allowed to examine the raw data from the Schmidt 1969 experiments. I hope that Schmidt (and Hyman) will agree to this suggestion.

Akers questions my concern about the generator bias problem in the Schmidt experiments, stating that Schmidt conducted extensive control runs and that even when these were cut into small segments, no bias was evident. My concern is that bias is not necessarily constant. Subjects were typically allowed to play with the machine and to switch into and out of "experimental mode" when they wished, allowing for the possibility of exploiting short-term biases, even if, over the long run, no bias is evident. As for the cutting into short segments, I presume that Akers is referring to Schmidt (1970a), where Schmidt cut the control runs into blocks of a size similar to test runs. He then used a goodness-of-fit approach to show that there was no bias. Unfortunately, this just tells us that there was no bias in the frequency distribution, based on short blocks. It does nothing to show that there were not, for example, short runs shortly after startup in which 4s, say, predominated and other short runs later on in which, say, 3s or 2s or 1s predominated. Overall, the frequency distribution by blocks can be unbiased, but so long as the subjects can opt in and opt out of the experimental series, they can still learn to exploit the underlying bias, if it exists. Furthermore, there may have been environmental differences between the times at which experimental and control runs were made (e.g., different loadings on power mains if test runs were run by day and control runs by night, as happened in some studies; differences even due to the presence or absence of human beings if the shielding was poor, and so on). With regard to the excess production of 4s noted in a precognition experiment (Schmidt 1969a), when subjects were supposed to be predicting, not influencing, the outcomes, it would only have been proper to balance the attempted production of 4s with the attempted production of 3s, 2s, and 1s, rather than using what was apparently the very same random-event generator in a later psychokinesis experiment (Schmidt and Pantas 1972), where psychokinetic success required that the subject influence the machine to produce 4s (not 3s or 2s or 1s).

In his commentary, Schmidt describes all sorts of correlations and so on regarding the control runs of his generator. This does not tell us that *at the time the subject was responding there was no bias*. Would it not be considerably easier just to listen to critics like Hyman and C. E. M. Hansel, and now J. Barnard Gilmore and Dawes, and do the studies in a way that eliminates this bias?

Fraud

I ignored the problem of fraud in my target article. However, it should not be totally overlooked, for although it is a problem in all research areas, it is potentially more serious in parapsychology given the lack of strong replicability. This has long been of major concern to Hansel. Broch argues that fraud is a *major* source of significant results in parapsychology, and even Akers, whom I know to be very conservative and respectful in his criticism, points to fraud as well and suggests that there is a strong possibility that psi arises primarily or entirely from a combination of experimenter error and fraud. Because the number of researchers claiming clearcut evidence is not large, he says, fraud would not have to be very widespread either.

Statistical Inference

Several commentators focused on the statistical evaluation of parapsychological data and the applicability of classical statistical inference. This is an important consideration, because the modern case for psi rests squarely on a statistical footing. Dawes's example shows vividly how the same data can lead to either significant or nonsignificant results depending on what one takes to be the unit of analysis. Gilmore has provided an excellent discussion of the problems associated with using "randomness" as a baseline against which a subject's guesses are compared. Gilmore also wisely questions the appropriateness of using classical statistics to evaluate data when the effect size is extremely small, and he urges parapsychologists to use the randomization test to eliminate this important concern. He also presents an appropriate experimental design for REG studies, which parapsychologists should consider very seriously. Dawes proposes a procedure for REG studies that is related to that of Gilmore in that, again, the calculation of exact probabilities is involved without making any guesses about the form of the distribution.

Obviously, parapsychologists attempt in each experiment to demonstrate the presence of psi, regardless of how psi is defined. Statistically significant deviations from chance are the usual indication that psi has occurred. If it were not that the presence or absence of psi by itself was very important to these researchers, there would be no need at all to discuss any of the literature that R and P cite because the effect sizes are so small. For example, if similar effect sizes were found in cancer research or in the search for a better glue, no one would pay any attention to them. It is in this vein that Clark Glymour draws attention to the fact that the p value is not a measure of the size of the effect. In general, the effects are miniscule in ESP experiments even though the p value is often extremely impressive. All this means, of course, is that one can be extremely sure that the null hypothesis is not true; but this gives no indication of *why* it is not true. (If all cars made by company X are exactly one micron longer than cars made by company Y, then the p value associated with rejecting the null hypothesis that the cars are of equal length would have an infinite string of zeros after the decimal point because the null hypothesis of no difference is absolutely false. In practical terms, however, such a difference would be unimportant.) Dominic V. Cicchetti presents a method for evaluating the magnitude of a psi effect and applies it to Schmidt's lamp-lighting experiment to show that the highly significant statistical effect is of utterly trivial consequence from a substantive point of view. Utts recommends the use of confidence intervals rather than exclusive reliance on p values, because a confidential interval communicates to independent readers just how large or small an effect is; this would be excellent advice for psychologists in their research as well.

Yet, of course, no matter how small, if there really is some paranormal effect, it is not insignificant from a philosophical and scientific point of view. On the other hand, as Navon says, because experimental controls can never guarantee that an experiment is not systematically biased, one should expect that any null hypothesis can be rejected with a large enough N, even if there is only a very small bias present. He contends that the smaller the effect, the more likely that it is due to an artifact.

Utts also argues that when effect sizes are small, very few studies should be expected to yield significant differences from chance, and that a typical ganzfeld study should be expected to obtain significant results only about one-third of the time, even if the true hit rate is 38 percent as opposed to a chance rate of 25 percent; she suggests that replications should not necessarily be expected to confirm the original results. This statement worries me because it could be taken to suggest that we should not be too concerned just because the replicability rate is low; the replicability rate is also certainly going to be low if there is no real effect.

Misleading the Reader

A small number of commentaries accuse me of misleading the reader (presumably deliberately) about parapsychological research. Tart mentions my "many misleading statements," but instead of addressing them, he also deals with two aspects of my apparent philosophical position. Gerd Hövelmann takes me to task for not pointing out that many of my criticisms of parapsychological research are to be found within the parapsychological literature itself. To the extent that this is true, and I do not disagree, I take it to be support for my position.

There are four specific criticisms about selective or biased reporting that require detailed rebuttal:

1. *The Hyman–Honorton exchange.* In my discussion of the Hyman–Honorton exchange, I quoted Hyman and Honorton (1986) regarding their conclusions about the database. In so doing, I left out three sentences, not in order to mislead, but in order *not* to mislead. These three sentences were as follows:

Although we probably still differ on the magnitude of the biases contributed by multiple testing, retrospective experiments, and the file drawer problem, we agree that the overall significance observed in these studies cannot reasonably be explained by *these* selective factors. *Something beyond selective reporting or inflated significance levels seems to be producing the nonchance outcomes.* Moreover, we agree that the significant outcomes have been produced by a number of different investigators. (p. 352, my italics)

Hövelmann and Roger D. Nelson and Dean I. Radin contend that those three sentences strongly contradict the general impression I convey about the status of parapsychological evidence. On the other hand, Spanos and de Groot fault R and P for *only* reporting the part that I left out, making it look as though Hyman came to agree that there were genuine anomalies in the ganzfeld data. Railton and Krippner both assert that both R and P and I have reported just those parts of the Hyman–Honorton conclusions that suit our purposes, omitting important aspects that do not.

I cannot speak for R and P, but my own selectivity was based on the following: Because nowhere in my target article did I address selective reporting, inflated significance levels, or the problems of retrospective experiments, I omitted the three sentences in question as completely irrelevant to my discussion. They refer to three specific sources of possible error, and to the fact that Hyman and Honorton agree that these three are not sufficient to account for the outcomes. However, Hyman's position is that various other flaws, and especially flaws in methodology, *could* account for those outcomes (see also Hyman's commentary):

If psi is responsible for the outcomes obtained in this data base, then the ganzfeld experiment should continue to produce successful outcomes *when the various problems that Hyman pointed out are eliminated.* (Hyman and Honorton 1986, p. 353, my italics)

2. *REG studies.* Both R and P and I had a section of our target articles dedicated to Schmidt's REG studies. Each discussed the early

Schmidt work (Schmidt 1969a) in some detail, and we both made mention of the most recent Schmidt study (Schmidt et al. 1986). Neither paper, in those sections, discussed other Schmidt research, except for a passing reference. However, Nelson and Radin single me out for focusing on a handful of Schmidt's early experiments and for failing to mention the many other random-event generator studies. They accuse me of ignoring their own body of research. I have carefully analyzed a considerable amount of the Princeton REG research (Part 2 of this volume). Because there was no space in the original target article for a detailed analysis, nor is there space in this reply, I can only say that I found as much reason to be concerned about the methodological soundness of experiments giving rise to that body of data as with the other evidence for the paranormal I have examined. For now, let me simply refer to Akers's advice that it would be wise to defer judgment on these REG experiments until more information is published about the experimental procedures.

With regard to Nelson and Radin's comments about my treatment of the May et al. (1980) report, again let me draw from the commentaries: Hyman contends that the same criticisms that May et al. levied against REG experiments carried out before 1979 continue to be applicable to the REG experiments that have appeared since.

3. *Child's criticism.* In his commentary, Child indicates that I twice mention his article (Child 1985) without revealing that one of its main points was that critics, including myself, have grossly distorted the most basic facts about the Maimonides research into ESP in dreams. The reader should examine Child's (1985) criticism of me, and the very little that I actually had to say on that subject (Alcock 1981) in order to evaluate the substance of Child's charge. As R and P point out, however, there have been no independent replications of this line of research that have provided significant results, and two major failures to replicate have been reported.

4. *The mediumistic evidence.* I am chided by Braude for avoiding a discussion of the evidence produced by "the best cases" of physical mediumship. Braude argues that these easily resist the traditional charges of error and fraud. This is a matter of opinion, of course, because

he is referring to events that took place many years ago with mediums D. D. Home, Eusapia Palladino, and others. He accuses me of invoking "one of the least impressive and most generally irrelevant cases of all—that of Uri Geller." That is very easy to say now, but I remind the reader that Geller was taken very seriously by many people inside and outside of parapsychology, and was subjected to considerably more critical scrutiny than people like Home and Palladino before his case became unimpressive and irrelevant. All in all, I believe Braude's argument should be directed not at me but at R and P, the advocates of parapsychology; they failed to make a strong case for the mediumistic evidence, if such a case can be made.

The Standards of Criticism

Criticism of parapsychology, argues Pinch, should *not* be equally applicable to mainstream science, and the critic should be able to specify the conditions under which the criticized work could avoid the criticism. I disagree with the first of these constraints, for my major criticism of parapsychology is that it offers neither theory nor replicable (in the "strong sense") effects. If there is some area of mainstream science where this is equally applicable, then I would level the same criticism in that direction. Pinch argues that I do not suggest how experiments could be improved. I did suggest a control procedure for Schmidt's REG studies, and if space permitted, I could make many other suggestions. However, in general, whenever flaws are found, the best strategy to improve the experiment is simply to get rid of those flaws!

Collaboration Between Proponent and Critic

To raise the level of debate between critic and proponent, Benassi urges both to "play in the same ball game." As an example of this, he refers to the Hyman–Honorton (1986) exchange in the *Journal of Parapsychology*. Whereas I too am pleased by that exchange, it remains to be seen whether it will have much impact in the long run on the way parapsychology is carried out. Pinch argues that bringing critics and proponents together to examine evidence is unlikely to resolve anything, and may actually exacerbate the debate, because they are unlikely to share crucial assumptions as to what constitutes a competently performed experiment.

Although I am generally sympathetic to Benassi's views, I am puzzled by his suggestion that if after a reasonable amount of time parapsychologists do not provide convincing data for communication anomalies then the scientific community should begin to ignore them. How long is reasonable? Over a century of formal empirical inquiry has been carried out so far.

Finally, I welcome Parker's invitation to collaborate in the design of parapsychological experiments. However, mutual supervision of the execution of such experiments is equally important, and I am not sure how practical that would be given that we live on different sides of the ocean.

In Conclusion

I continue to believe, along with Hyman (and apparently also Palmer, judging by his commentary), that the existence of psi anomalies, let alone the paranormal, has not yet been demonstrated. To those who are new to this debate and do not know whom to believe, I suggest that rather than listening to an acknowledged skeptic like myself, or to acknowledged proponents like R and P, they reread the commentaries of parapsychologists like Akers and Blackmore who have developed their skepticism from *within* parapsychology rather than from the outside looking in.

In my view, parapsychology cannot have survived for more than a century on the diet of substantiated research findings and theoretical advances that sustains research in other fields, because it has not been blessed with any of these. In order to avoid being misunderstood a second time, I shall eschew my soul and dualism metaphors and argue simply that parapsychology is not anomaly driven, but represents a quest to demonstrate that the materialistic worldview that predominates in modern science is incomplete, and that personality/mind/

thought can interact directly with matter. This, *by itself*, does not make parapsychology unscientific. Of course, if the mind–matter interaction can be demonstrated, then, as Joseph Banks Rhine (1943) pointed out, there is at least some reason to believe that the human personality may not be tied to the fate of the body.

Because, I, too, am interested in experiences that seem anomalous, I totally agree with Nadon and Kihlstrom, who stress that future research needs to elucidate the cognitive nature of anomalous experiences and to examine the situational and dispositional factors involved in their occurrence. As Blackmore (1983a) has counseled, let us get on with the study of anomalous experience and leave the psi hypothesis aside for now, for it only gets in the way. This entire exchange of views, as profitable as it has been for me personally, has had little or nothing to do with anomalous experience, and that is because the focus is not on what people think or feel or experience, but on whether or not psi, presumed by parapsychologists to be very relevant to such experience, can be demonstrated to exist. I believe that that indictment applies to parapsychology in general. I agree with Truzzi that inquiry should not be blocked. However, the psi hypothesis is likely, ironically, to hinder progress in understanding the very experiences that parapsychologists say they want to explain.

Note

1. The writers I mention specifically are: Victor G. Adamenko of the Parapsychological Association; Charles Akers, Newtonville, Mass.; Gary Bauslaugh, vice-president of instruction, Malaspina College, Nanaimo, B.C., Canada; O. Costa de Beauregard, Institut Henri Poincare, Paris, France; John Beloff, Dept. of Psychology, University of Edinburgh, Edinburgh, Scotland; Victor A. Benassi, Dept. of Psychology, University of New Hampshire; Barry L. Beyerstein, Brain Behavior Laboratory, Dept. of Psychology, Simon Fraser University, Burnaby, B.C., Canada; Susan J. Blackmore, Brain and Perception Laboratory, University of Bristol, Bristol, England; Stephen E. Braude, Dept. of Philosophy, University of Maryland, Baltimore; Henri Broch, Laboratory of Biophysics, University of Nice, Nice, France; Mario Bunge, Foundations and Philosophy of Science Unit, McGill University, Montreal, Canada; Irvin L. Child, Dept. of Psychology, Yale University; Domenic V. Cicchetti, Veterans Administration Medical Center, West Haven, Conn.; Robyn M. Dawes, Dept. of Social and Decision

Parapsychology: Science of the Anomalous or Search for the Soul?

Sciences, Carnegie-Mellon University, Pittsburgh, Pa.; D. C. Donderi, Dept. of Psychology, McGill University, Montreal Canada; Antony Flew, Social Philosophy and Policy Center, Bowling Green State University, Bowling Green, Ohio; Martin Gardner, Hendersonville, N.C.; J. Barnard Gilmore, Dept. of Psychology, University of Toronto, Toronto, Canada; Clark Glymour, Dept. of Philosophy, Carnegie-Mellon University, Pittsburgh, Pa.; C. E. M. Hansel, Dept. of Psychology, University College of Swansea, University of Wales, Swansea, U.K.; Ray Hyman, Dept. of Psychology, University of Oregon, Eugene, Ore.; Brian D. Josephson, Dept. of Physics, University of Cambridge, Cambridge, England; Stanley Krippner, Dept. of Psychology, Saybrook Institute, San Francisco, Calif.; Brian Mackenzie, Dept. of Psychology, University of Tasmania, Hobart, Tasmania, Australia; Robert Nadon, Dept. of Psychology, Concordia University, Montreal, Canada, and John F. Kihlstrom, Dept. of Psychology, University of Arizona, Tucson, Ariz.; David Navon, Dept. of Psychology, University of Haifa, Haifa, Israel; Roger D. Nelson, School of Engineering/Applied Science, and Dean I. Radin, Dept. of Psychology, Princeton University; John Palmer, Institute for Parapsychology, Durham, N.C.; Adrian Parker, Dept. of Child and Youth Psychiatry, University of Göteborg, Göteburg, Sweden; Trevor Pinch, Dept. of Sociology, University of York, Haslington, York, England; Peter Railton, Dept. of Philosophy, University of Michigan, Ann Arbor, Mich.; John T. Sanders, Philosophy Committee, Rochester Institute of Technology, Rochester, N.Y.; Helmut Schmidt, Mind Science Foundation, San Antonio, Tex.; Nicholas P. Spanos and Hans de Groot, Dept. of Psychology, Ottawa, Canada; Rex G. Stanford, Dept. of Psychology, St. John's University, Jamaica, N.Y.; Charles T. Tart, Dept. of Psychology, University of California, Davis, Calif.; Jerome J. Tobacyk, Dept. of Behavioral Sciences, Louisiana Tech University, Ruston, La.; Marcello Truzzi, Dept. of Sociology, Eastern Michigan University, Ypsilanti, Mich.; Jessica Utts, Division of Statistics, University of California, Long Beach, Calif.

PART 2

Psi in the Laboratory

Introduction

It was Joseph Banks Rhine who introduced parapsychologists to the laboratory. Although a botanist by training, after obtaining his doctorate Rhine devoted his entire career to parapsychology. As a university student, he had been troubled by the conflict between, on the one hand, the religious beliefs with which he had been raised (he had originally planned to become a minister) and, on the other, the skepticism about these beliefs engendered by his training in science. As did many other pioneers in parapsychology, Rhine considered the scientific study of putatively psychic phenomena to be a good compromise between his respect for science and his conviction that there is more to our existence than materialistic philosophy suggests.

In the 1920s, around the time of his graduation from university, Rhine contacted and found favor with William McDougall, a leading social psychologist who was also unable to accept strict materialism and was interested in scientific investigation of the psychic realm, a realm which, through the popularity of spiritualism in the late nineteenth and early twentieth centuries, certainly seemed to cry out for investigation. (In fact, mainstream psychologists on both sides of the Atlantic had given serious consideration to spiritualists' claims, but had become disillusioned and had abandoned the quest because of the failure to find anything other than fraud.) When McDougall accepted the post of chairman of the psychology department at Duke University, he invited Rhine to work under his supervision in the investigation of a body of transcripts of mediumistic communications, and this grew into a full-time position, with Rhine eventually setting up his famous Parapsychology Laboratory.

It is not surprising that, finding himself in a psychology department and under the guidance of a psychologist, Rhine adopted an experimental psychological approach to the study of psi. As psychologists had done in their studies of "normal" human behaviors and abilities, Rhine came to depend almost exclusively upon statistical analysis as the arbiter of whether or not anything psychic was going on in his studies.

Rhine built his quest for empirical evidence of psi on what, for want of a better term, might be called the "wishing-guessing" paradigm. In the study of psychokinesis, for example, a subject would watch dice being rolled by a machine and would "wish" the dice to come up in some prespecified way. Rhine would calculate the success rate over a (usually very large) number of trials and then, by means of statistical tests, would decide whether or not the observed success rate exceeded that expected if only "chance" were operating. In the study of precognition or telepathy or clairvoyance, the typical task involved a set of 25 cards consisting of 5 sets of 5 different symbols (the "Zener" deck). The subject's task would be to guess, as each of the cards was placed face-down on the table, which of the five symbols was on the target card. The wishing/guessing paradigm was an attractive one in that it made the likelihood of the event occurring by chance readily calculable.

Because of Rhine's claims of above-chance scoring, there ensued in the 1930s and 1940s considerable debate between Rhine and his defenders and a number of psychologists who challenged the validity of Rhine's statistical analyses. Although some of the attacks struck home, leading to changes in the ways that parapsychologists did things thereafter, the brunt of the assault on statistical grounds was successfully deflected.

Although Rhine came to believe that he had clearly demonstrated the reality of psi through his card-guessing and dice-rolling studies, most scientists refused to accept that claim because of the lack of replicability of Rhine's demonstrations and their belief that his research was flawed by his failure to institute adequate experimental controls. With regard to dice-rolling, for example, Rhine himself found that as controls were made more rigorous, PK effects tended to disappear. Evidence from dice-rolling studies is no longer held in much esteem by most parapsychologists.

There are a small number of "classic" Rhine studies that are still sometimes adduced as evidence by parapsychologists, but which, because of the lack of replicability, stand essentially as reports of miracles; some may believe that psi effects really were produced, while others may well consider that the results were due to some sort of methodological artifact that may not be obvious from the written reports, and still others may even posit fraud as the explanation. While one must be wary when pointing the finger of fraud, and while it must be emphasized that there has never been any evidence to suggest that Rhine cheated in any way in his research, the possibility of fraud should never be overlooked completely. Some important parapsychological demonstrations by one of Rhine's leading contemporaries, S. G. Soal, fell into disrepute as evidence grew that Soal cheated.

Psychokinesis Research

The history of psychokinesis research is especially relevant here since the random-event generator research, which will be discussed in detail, falls in that domain. Stanford (1977) described three phases of modern PK research. The first of these, from 1934-1950, he labeled the "early quantitative-experimental period." It was dominated by the work of Rhine and his colleagues and, as mentioned earlier, dice-rolling was the major experimental PK task. A subject would attempt to influence a die in motion so as to make it stop with a particular face up. However, as Stanford points out, the methodology was often less than rigorous in these early experiments; dice were often hand thrown or cup thrown, and each of these methods is subject to bias. Moreover, there was the problem of bias in the die itself: Due to the fact that the higher faces of the die are lighter, they are likely to turn up more often. This problem was not corrected, as it should have been, by balancing the target faces across trials.

In about 1944, the "quartile decline" (QD) effect was discovered. It was found that PK success was unequally distributed over the period of testing and that there was a typical pattern in the success rate: If one divided the results for a session in four quarters, the success rate for the first quarter was higher than that for the last one. In a reanalysis of 18 studies by Rhine and his colleagues up to 1943, the vast majority showed this effect. However, Stanford points out that, while the decline effect and other similar "internal" effects occasioned considerable interest because they seemed impervious to methodological artifact, die bias and a number of other factors having nothing to do with psi *can* produce such effects. At any rate, from 1944 to 1951, more research was carried out that looked at internal effects, explored new testing methods, and attempted to overcome known methodological problems.

Stanford refers to the period from 1951 to 1969 as the "middle period." Here the die face method fell into relative disuse as the "placement" method superseded it to a large degree. With this method, the subject attempts to influence an object, such as a die or a ball, to move in one direction or another during its roll. However, this methodology, as had die-rolling, failed to yield convincing data. As Stanford commented (p. 238):

> The 1960s evinced a clear decrease in the number of PK studies done and reported. Some investigators seemed to feel that PK results were difficult to get and were weaker and less reliable than in the case of ESP.

In Stanford's view, contemporary PK research has been revolutionized by the introduction of electronic random-event generators (REGs). Such apparatus appeared to answer Rhine's critics, such as

C. E. M. Hansel, who had called for the use of automated equipment that would both produce the targets at random and record and analyze the scores. The use of such equipment would mean that one could be reasonably certain that any observed departures from chance were not due to biases in the target sequences, or to errors in recording or analyzing the data, or to "sensory leakage" (i.e., the transmission, consciously or unconsciously, of information about a target from a sender to a receiver by normal sensory channels).

Parapsychology and Quantum Mechanics

When a radioactive material decays, it takes a quantum leap from one energy level to a lower one, with the decrease being manifested by the emission of a subatomic particle (or ray). Such decay and particle emission is considered by quantum mechanical theory to be a *truly* random process, in that there is in principle no way to predict when such an event will occur. Such a process, then, provides a perfect source of randomness for a random-number generator, and all one has to do is to set up some cyclical electronic process (for example, repetitively cycling through the digits 1, 2, 3, and 4) that is stopped by the quantum emission, yielding a random number. In principle, one could use any range of numbers. Usually, in parapsychological research, the number of alternatives is limited, with a binary output (two possible numbers) or four outputs being the most common.

Such equipment seems naturally suited to the study of ESP, where it is important to generate a target series that is unique and unbiased; but how can the equivalent of a random-event generator be used to study the ability of the mind to directly influence matter? The answer to this is to direct PK toward the random-event generator itself; if the generator output is clearly nonrandom while the individual is attempting to influence it, but is random at other times, then this would seem to support the PK hypothesis. Indeed, if PK does exist, it would appear an easier task for the human mind to influence an electronic or subatomic process than to deflect a die in mid-flight. However, with the die, at least one sees the target and understands the task. With a random-event generator driven by radioactive decay, it is difficult to imagine how one would identify which atom is about to decay next in order to postpone it for a brief interval, and it is also difficult to imagine what form the wish might take. While one might well imagine that subjects in a die-rolling study in which the goal is to produce sixes might be repeating over and over to themselves "Come on, six, come on," what would one wish for in the quantum case? Would it be "Don't jump, don't jump"? (As we shall see, some parapsychologists have more or less determined that one need not know anything at all about what is actually going on at the quantum level—all that is necessary is that the subjects work their wish upon the ultimate manifestation of that quantum process, be it the lighting of a particular lamp from among a set of lamps, or whatever.

Not only does the use of a radioactive source as the basis of randomness provide true randomness for psi studies, but once one directs one's thinking to the level of quantum mechanics it is natural, especially when one is trained in physics, to think about the possible connection between alleged psychical phenomena and some of the paradoxes about reality that are served up by quantum mechanics. Quantum theory paints a new and dramatically different picture of reality. As Helmut Schmidt (1979c, p. 208) points out, in contrasting the world as it was understood to be in the days of an earlier champion of parapsychology, Charles Richet (a preeminent physiologist and a winner of the Nobel Prize in 1913), with the world as seen by modern physics:

> Richet based his discussion of precognition on a deterministic model in which the future could, in principle, be calculated from the present. Modern quantum theory, on the other hand, suggests that the future is not completely determined by the past and that there are processes, the quantum jumps, which are, in principle, unpredictable. Therefore, the physicist's most basic question with regard to precognition is whether human subjects can predict the outcome of quantum processes, like radioactive decay.

Indeed, almost all of the work that has been carried out in the areas under review here (i.e., the ability to influence random-number generators and the ability of a subject in the laboratory to describe distant sites that are being visited by a "sender," which is referred to as "remote viewing") has been done either directly by, or under the supervision of, physicists or engineers. This is not to suggest that parapsychology (or "paraphysics," as some physicists who interest themselves in paranormal processes would have it) has been accepted as a legitimate area of research by these disciplines. On the contrary, only a very small number of people are involved, and in many cases they are viewed by their contemporaries with suspicion or outright disdain. Some of them have abandoned their original discipline, while others are careful to keep their interest in paranormal processes quite separate from their more orthodox research.

The study of the mind's putative ability to influence quantum processes was pioneered largely by Helmut Schmidt, a physicist who now dedicates himself to the study of the paranormal. Robert Jahn, Dean of Engineering at Princeton University, is also carrying out an extensive research program in the same area, although he does not consider himself to be a parapsychologist or paraphysicist. As for remote viewing, the development of this area in parapsychology was carried out at Stanford Research International (SRI, formerly the Stanford Research Institute) by Russell Targ and Harold Puthoff, both former laser physicists.

Scope of the Review

In an area as controversial as parapsychology, there is often disagreement about what evidence or which research reports really reflect the mainstream of the domain. Critics are sometimes accused of holding up the poorer examples of parapsychological research as though they were representative of the best the area has to offer. In order to avoid the possibility of such charges, and in order as well to make the reviewing task more straightforward and manageable, this present review is limited to those research papers already selected for review by a leading parapsychologist, John Palmer (1985c). His paper, prepared for the U.S. Army Research Institute for the Behavioral and Social Sciences, evaluates eight areas of parapsychology, including the two this review addresses. Since he has already selected the studies that provide the best case for remote viewing and for mental influence on random-event generators (and I agree with his delineation in this regard), there should me no apprehension that the studies reviewed herein reflect the bias of a skeptical reviewer. (I have added a few more recent papers by the same authors Palmer chose; I am sure that these would have been included in Palmer's review had they been available at the time. Their inclusion in no way changes the outcome of my evaluation.)

In reviewing the various research reports, the guiding principles shall be those that one would use in reviewing a submission to a psychology journal: Is the experimental design adequate and is the control of possible extraneous variables stringent enough? Is the statistical analysis appropriate? Are the conclusions justified by the procedures and data?

The question of possible experimenter fraud will not be directly addressed, for it seems to this writer that there is little point in basing the accusation of fraud on a written report. If one wants to cheat, and if one has a normal portion of intelligence, then one would write the report in such a way as to make the cheating impossible to detect. It is only through analysis of raw data and other extra-report aspects that one can really detect chicanery.

1
Parapsychological Research Using Random-Event Generators

The Work of Helmut Schmidt

Over the past eighteen years or so, Helmut Schmidt, has promoted the use of electronic random-event generators in parapsychological research. As John Palmer (1985c) has observed, Schmidt's psi research has passed through several rather distinct stages. In the beginning, his investigations bore primarily on the question of whether or not psi exists. Using a modulus-4 random-number generator, Schmidt would have his subjects press one of four buttons, each under a lamp, and then depending on the output of his generator, which was ultimately dependent on radioactive decay, one or another of the four lamps would light up. Schmidt was able to study precognitive ability as well as psychokinesis (PK), although he admitted that it was not possible to totally exclude one or the other process from any given experiment.

In the next phase, Schmidt moved away from the study of precognition and turned to the exclusive study of PK. Typically, he employed only two targets instead of four, and a button press initiated a short series of trials rather than a single trial. He employed various forms of feedback, including a circular display of nine lamps and a series of clicks delivered to the subject by earphones.

The third phase began with the publication of Schmidt's quantum mechanical model of psi (Schmidt 1975), which, based on the notion that some microscopic events are not determined until they are observed, suggests that subjects can employ PK to influence events that occurred in the past but have not yet been observed. This theory led him to shift his research in the direction of the investigation of psi effects on previously generated and recorded random series.

In recent years, Schmidt has shifted his focus again. While continuing to explore the ostensible effects of PK on prerecorded series of random events, instead of using an actual series of random events he now uses randomly generated seed numbers, which when fed into an algorithm will generate a final score. Again, the subject's task is to alter the series in some way.

After carefully reviewing the database of the Schmidt publications used by Palmer (1985c) in his review, I am in agreement with Palmer that one can hardly explain away Schmidt's results in terms of chance occurrence. Almost all of the fifteen research papers under consideration (the fourteen reviewed by Palmer plus one more recent one [Schmidt 1985]), yielded substantial p-values. While Palmer has gone to considerable lengths to estimate an overall z-score for the combined results from the fourteen papers (which he finds to be $z = 9.92, p < 10^{-12}$), I have no great confidence in statistics based on the conglomeration of a group of diverse studies, and I am content simply to say that Schmidt has accumulated some pretty impressive evidence that something other than chance is influencing the subjects' scores.

What might this influence be? One explanation, that preferred by Schmidt and most other parapsychologists, is that the influence is a psychic one, a psi effect brought about either by the subjects themselves, or as Palmer (1985) discusses, perhaps by the experimenter, Schmidt himself. On the other hand, some critics, notably C. E. M. Hansel (1980), have leaned more toward an explanation based on fraudulence on the part of the experimenter.

For my part, I find both of these interpretations extreme because both make assumptions that cannot be backed up simply by looking at the data or the experimental reports. To say that one has evidence of psi just because scoring occurs at rates significantly above or below chance over a number of studies is jumping to a conclusion, and to argue that the only way in which such extra-chance scores could have come about is through fraud is to make a similar leap.

My own lengthy analysis of Schmidt's papers (see Appendix 1) leads me to respect Schmidt for what I see as an honest effort to improve the quality of his research over the years. I am also impressed by his creativity; some of his experiments border on the ingenius in some respects, although the ingeniousness is often badly tarnished by unnecessary complexity and by weak methodology.

That being said, it is also my strong opinion that, with very few exceptions, Schmidt's studies are seriously flawed to the extent that there is no way of knowing whether the data are "anomalous" in some way or simply the result of the lack of both empirical rigor and good laboratory housekeeping. Each of the major flaws, along with some general criticisms of Schmidt's research, are presented below. A more detailed treatment of each of the research articles can be found in Appendix 1.

General Criticisms of Schmidt's Work

Disjointedness. Despite the threads that link various studies to succeeding ones, there is a lack of systematic inquiry evident in Schmidt's work. One study does not usually follow logically from another, and Schmidt neglects to do any in-depth probing of factors other than psi that might have generated the obtained departures from chance in his experiments. Each experiment, or at least each group of two or three studies within a given research paper, tends to stand as an independent "miracle" in a sense, and having produced the miracles Schmidt moves on without really "nailing down" just what it is that was going on.

The lack of coherence in Schmidt's research thrust is perhaps most evident by the fact that he switches from one type of random-event generator to another and from one type of task to another so frequently (often on an experiment-to-experiment basis) that it is impossible to really get to "know" his generator (see Table 1). Indeed, the changes from study to study go well beyond what is suggested by Table 1, since the methodology changes frequently even when the source of true randomness stays unchanged.

As Ray Hyman (1981) has pointed out, were Schmidt to use one REG over and over, he would allow himself and others the opportunity to come to understand the peculiar properties of that particular generator, whatever they may be, and be in a position to debug it. As it is, we are expected to accept his word for it that there is no bias in the generator itself.

Inadequate controls. There is a general disregard for experimental control running throughout most of Schmidt's experiments, much of which is pointed out in the detailed analyses of his studies given in Appendix 1. Schmidt seems to make the unacceptable assumption that instrumentation can replace old-fashioned controls in human experimentation (Hyman 1981). Schmidt's research rarely involves any kind of control group. For example, it would be germane to compare a group of subjects who are both allowed play sessions and allowed to decide when to start the test runs with a group allowed play sessions but for whom the test runs begin at prespecified times.

It would also be useful to run trials on which there is no feedback for comparison with those for which there is feedback. Without feedback, the subject would have a difficult time keying in to short-term biases. Schmidt may well argue that the lack of feedback weakens the likelihood of psi either because of lessened motivation or interference with goal-directedness, or even because *observation* is essential (this according to his 1975 theory). (The notion that the subject needs normal, sensory feedback in order to be able to motivate or guide his or her paranormal sensory or kinetic abilities seems a bit odd, however. If extrasensory processes can reach into the future or down to the subatomic level, surely they might just as readily provide feedback

TABLE 1

Overview of Studies Included in Schmidt Review

Study	Source of Randomness	Task/Goal/Circumstances
1969a	modulus-4 quantum REG	precognition or PK
1969b	Rand tables	precognition or clairvoyance
1970a	binary quantum REG	PK
1970c	binary quantum REG	cat, cockroaches as subjects
*1972	modulus-4 quantum REG	internally different machine
1973	electronic-noise REG	targets at high speeds, visual vs auditory feedback
1974	two generators, one simple, one complex	examination of role of generator on PK performance
1976	electronic noise	prerecorded targets
1978	binary REG	prerecorded targets
†1978	indeterminate REG—an IMSAI minicomputer	prerecorded targets
1979a	binary REG	PK in cockroaches
1979b	electronic die based on radioactive decay	prerecorded and real-time events, stroboscopic light as reward
1981	seed numbers produced by binary quantum generator (modulo-16)	prerecorded and pre-inspected seed numbers
1985	computer with attached Geiger counter	effects of two successive PK attempts
1986	computer with attached Geiger counter	channeling PK

* With Pantas.
† With Terry.

about hits and misses as well.)

Some of Schmidt's more recent studies did involve a control series. Terry and Schmidt (1978) included a control series of targets, but this control series turned out to show evidence of bias!

Lack of data-snooping. Schmidt consistently fails to do the sorts of post-hoc data-snooping that one would expect in the face of findings of the kind he has produced. One would think it essential to examine the actual target series used in each of his experiments, even if one is prepared to accept that any departures from randomness are caused by PK. If, for example, there is an excess of 4s in a target series using the modulus-4 generator, it would be important to try to repeat the study and produce an excess of 3s or 2s or 1s. If only 4s ever appear in excess, this obvious bias should be evident even to those who believe in PK. The point is that by snooping around one might find valuable clues in the target sequence that would lead the way to a source of bias.

Experimenter isolation. Schmidt has worked in relative isolation from other experimenters, and this of course makes it more difficult to evaluate his work. Except for the 1986 study, his raw data is generally unavailable to other researchers and critics.

Ad-hoc hypotheses and experimental goals. Schmidt, in some of his studies, sets up experimental manipulations that are so complicated and contrived that one might suspect that in some instances part of his description of the experimental manipulation came into being after the data were examined. (This is not an accusation but merely an allusion to the possibility that, as occasionally happens in psychology experiments, the less than rigorous researcher persuades himself or herself after the fact that the subjects were really out to obtain results different from those *originally* targeted.)

For example, in Schmidt (1970a), the subjects were encouraged to think pessimistically and in terms of failure. Yet Schmidt alluded in this paper to the notion that PK is goal-oriented, that even in a complicated set of circumstances, results are obtained by concentrating only on the goal. Here the goal was self-contradictory: Subjects were supposed to try to influence the lamps to illuminate successively in the direction of their choice, but they were also supposed to want to fail! Why not have them concentrate on having the sequence go *opposite* to their preferred direction? It was actually worse than that; if the subject chose to try to make the lamps light in a counterclockwise direction, a switch was flipped to cause a +1 number from the binary REG to move the illumination of the lamps in counterclockwise direction, so that failure (an excess of –1s), which is *really* success (because subjects are encouraged to psi-miss), is now linked with perceived success on the board, whereas when the subject chose the clockwise direction, failure (an excess of –1s, which again is really success) is associated with perceived failure on the board. What is the goal-directed PK going to do?

Another example is that of Schmidt and Pantas (1972), where subjects were instructed to try to *psi-miss* the number 4 outcome of a modulus-4 REG; yet the attempt is also made to discourage them about their ability to succeed by such failure, to the extent that they end up failing to psi-miss, which is manifested by psi-hitting by generating an excess of 4s.

Randomness. Since almost all of Schmidt's work requires that observed scores be compared with a "chance" score, it is critical to Schmidt's interpretation of his results to be able to assume that any bias in the data came about *after* the original generation of such data.

It is to be expected that when a subject is trying to predict which of, say, four lights is going to be "randomly" chosen to light up next, and when there is a bias in the target sequence, the subject might quickly learn to match his or her response frequencies to the target frequencies, thus leading to an increased hit rate suggestive of precognition. The psychological literature, of which Schmidt, a physicist, seems to be woefully unaware, contains considerable evidence that human beings, in almost precisely the sort of situation that Schmidt has so often used, can quickly learn to match the frequency of their responses to the frequency of the diverse targets. This is referred to as "probability learning": relative frequency judgments tend to match the actual relative frequencies in such experiments (e.g., Estes 1976; Radtke, Jacoby, and Goedel 1971).

In the PK situation as well, if there is a short-term bias in the target series, we might expect that subjects will be able, at least in some cases, and depending on the degree of bias, to detect it. For example, in the case of Schmidt's modulus-4 generator, if one light lights up more frequently than the others, the subject might then direct more guesses toward that light, thus bringing about a higher hit rate. This criticism applies to virtually all of Schmidt's studies in which subjects actually have to do something specific (i.e., as opposed to the retroactive PK studies where subjects really did nothing but "wish").

Indeed, using the modulus-4 generator and four lamps in the early precognition study, one subject reported that instead of using precognition, he had attempted to use PK to produce more red lamp lightings, and sure enough, subsequent analysis of the target sequence indicated an excess of 4s, which gave rise to reds. Subsequently, in the PK study mentioned earlier using the same apparatus (Schmidt and Pantas 1972), in which double psi-missing produced an excess of 4s, a hit occurred only when a 4 was generated. Obviously, if the apparatus was producing more 4s than it should have, on a short-term basis perhaps, the subject in the first study may well have detected this and believed that he was causing it, and the same bias could account for significant results in the subsequent study. Schmidt (1976) also referred to a pilot study by Lee Pantas using presumably the same modulus-4 generator, and again the subject's task was to produce an excess of 4s, which the subject succeeded in doing! Schmidt (1976) also described another, similar study, using the same test situation, carried out by E. F. Kelly working with the subject Bill Delmore (who has achieved some fame and/or notoriety in parapsychological circles). Again a significant excess of 4s was produced. One might think that the careful researcher might want to try the same approach with 3s or 2s or 1s as targets. There is certainly good reason to be suspicious of that modulus-4 REG.

Although Schmidt's later work shows greater sophistication in some ways, involving two random determinations rather than one, he continues to ignore or be unaware of suggestions that would surmount some of the concerns about the generator. Hansel (1980), for example, suggested that *pairs* of runs be generated, and for each pair, one run be designated the experimental run and the other the control run, on the basis of some random process, such as a coin toss. He also urged that the experimenter be kept blind as to the nature of each run. One could extend this down to the actual level of trials within a run: for each trial, one could take the next two generated targets, and assign, using another random process, one as the target and the other as a control. Then, at the end, one would have both a target series, modified perhaps by the PK of the subject, and a control series against which one could more properly evaluate the effect of the subject's attempted intervention. If there were biases introduced into the generation process, these would be as likely to reflect themselves in the control series as in the target series, and thus one could rest comfortably in the knowledge that any deviations from randomness in the test series could be evaluated in light of the control series. Schmidt may of course be concerned about the possibility of the subject's PK "spreading" ("displacement effect") to influence both elements of the pair of random numbers. However, since he is persuaded that the subject does not direct his PK toward the underlying event, but rather simply concentrates on obtaining the macrolevel result that he is seeking, one would think that he need not harbor any such concerns.

If one examines the fifteen reports in detail, one finds that in about half of them there is no mention of randomization controls at all. In many of the other reports, Schmidt argues that the random generators have been demonstrated to be unbiased on the basis that he has run long control series that are demonstably free of bias. (See Appendix 2 for a detailed list of the randomization test procedures used in the various studies.)

Following is a list of problems having to do with randomization checks; the numbers corresponding to these problems are used in the "Randomization Checks" of Table 2 to indicate the studies in

TABLE 2

Major Problems, Schmidt's Confirmatory Studies

Study	Randomization Checks	Methodological Lack of Rigor	Recording Problems	Security
1969a-1	1, 3, 4	2, 3b, 3c, 4, 5a, 5b, 9	2	1
-2	1, 3, 4	2, 5a, 5b, 9	1, 2	1
1969b	Unspecified‡	2, 5a, 5b, 6	1, 2	
1970a	1, 3, 5	2, 7	1, 2**	
1970c-1	1, 2	3a	2	
-2	1, 2		2**	
*1972-1	Unspecified	4, 7, 9	1, 2	
-2	Unspecified	1, 9	1, 2**	1
1973	1, 2	1, 2	2**	
1974	Unspecified	1, 2, 4, 5b	2**	2
1976-1	1			
-2	1			
-3	1			
1978-1	Unspecified	10		
1978-2	Unspecified	10		
†1978-1	1			
-2	1			
-3	1			
1979a	Unspecified	11		
1979b	Unspecified	2, 5, 10		
1981	Unspecified	2, 3c, 5		
1985	Unspecified	1		
1986	Not needed	1, 2, 8		

* With Pantas.
† With Terry.
** Output on punched paper tape.
‡ Reader is referred to unpublished paper for details.

which these weaknesses are found.

1. The test sequence is typically very much shorter than the control run. While the long control series may not demonstrate any bias, if we were to look at a series of targets as short as the test series, such bias may be evident. (Short runs will have a distribution that has a greater variance than will long runs, thus producing more deviations).

2. In the earlier studies particularly, the long control runs (randomization checks) were often run overnight or at other times when no one was about. This means that the machine was allowed to operate in a quiescent state, undisturbed by the voltage fluctuations that might result from the plugging in of electrical instruments elsewhere on the same voltage feed, which are likely to occur much more frequently during the daytime working hours, and undisturbed as well by the presence of human beings. (Equipment that is improperly shielded may even be influenced by the movements of a subject seated near it. If such were the case with Schmidt's generators, subjects could learn without awareness to make movements that tend to have the consequence of increasing or decreasing their scores, as required.) Also, if there were a warm-up effect (i.e., if the generator produced biased outputs until it was fully "warmed up"), then this effect would wash out over the long control runs, but could play a significant part in the short test-runs.

3. Often there is not enough information in the written report to indicate the temporal relationship of the control runs to the test runs.

4. While not relevant to the use of a binary REG, one should be careful when using an REG like Schmidt's modulus-4 to look not only for bias in terms of the distribution of the various outcomes but also for higher-order biases, such as doublets, triplets, and so on. Thus, if a 3 follows a 1-4 sequence more often than it should were there no bias, subjects could learn this partial contingency and use it to increase their hit rates. Schmidt never checked beyond the doublet level when using his modulus-4 generator.

5. In one study (Schmidt 1970a), Schmidt attempted to correct

for the difference in size between control runs and test runs by analyzing the control run data in blocks of a size similar to the test runs. He found, using a goodness-of-fit approach, that there was no bias, even using the shorter blocks. However, this approach fails to eliminate the problem, for all it does is to look at a frequency distribution, rather than examining the blocks themselves for independence from their neighbors. For example, the generator could produce above-average scores for the first fifteen minutes of its operation, and then settle down to slightly below average scores for another fifteen minutes.

Sometimes, as in Schmidt (1970a), the outputs of the generator were reversed halfway through the experiment, and it was stated that if there was a systematic bias, this would compensate. However, no data are provided for the target sequence before and after this change, and so we have no way of knowing whether there may have been a start-up effect, or whatever, that may have influenced the overall scores despite the change in the outputs. Such a procedure in any case only corrects for a constant bias and not a fluctuating one.

Data pooling. In his data analyses, Schmidt typically pools the data from all the subjects. One of the three subjects in the first experiment did not score significantly higher than chance, but his data were put together with the others to yield overall significance. The problem with this approach is that if one subject, for whatever reason, were to score very high—and this might in some instances be because of methodological artifact or even fraud—then the pooling of data might yield overall significance, whereas it might be more reasonable to point out, for example, that one subject scored remarkably well, while others did not. My objections here are perhaps picayune, but I must ask this question: Why use a number of subjects if indeed they are interchangeable? Why not do the whole study with one subject? Indeed, as will be seen later, on occasion Schmidt does exactly that.

Lack of methodological rigor: Most of the Schmidt studies suffer from a lack of methodological rigor. He often departs from what an experimental psychologist would consider to be standard operating procedure. There is no good reason for his failure to follow sound research practices, even while in most instances this failure is unlikely to have caused any real harm. However, in some instances the problem is rather serious and indeed may well be the source of his obtained deviations from chance.

Following is a list of methodological weaknesses found in Schmidt's studies; these weaknesses are referred to by number in Table 2.

1. In a number of the studies, Schmidt serves as both experimenter and subject; in at least one case, he is really the sole subject. In another (Schmidt and Pantas, 1972, second study), Pantas is the sole subject. This is a clear violation of sound research practice.

2. In many of the studies, there are varying numbers of trials and/or sessions per subject. Schmidt views his studies as attempts to score above chance, and it makes no difference in his mind if one subject contributes 3 trials and another 30. While the differing numbers of trials and sessions is not a major worry, nonetheless it does lead to discomfort, for if one subject is particularly good at detecting generator bias, and if for whatever reason that subject takes the lion's share of the trials in a given study, overall significance could result from that sole subject's scoring.

3. (*a*) In some of Schmidt's earlier studies, the number of trials and/or sessions was not specified in advance, and this of course allows for optional stopping. (*b*) Sometimes a range was prespecified, for whatever reason, and again this allows the experimenter to stop the study within that range at a point where the results tend to confirm his hypothesis. Given the long debates in parapsychology about the optional-stopping problem, one wonders why a parapsychological researcher would allow himself to build such optional stopping into his procedure, even if it can be shown that it would not affect the data very much.

4. In some of the studies, the actual number of trials falls considerably outside the prespecified range. For example, in Schmidt (1969a, experiment 1), each of the three subjects was assigned a range of trials. One finds upon examination of the data that the subject whose range was 15,000 to 20,000 trials actually completed 22,569

trials, while one of the subjects who was supposed to carry out 20,000 to 25,000 trials actually carried out only 16,250 trials! There is no excuse for such sloppiness, but at least it is a tribute to Schmidt's honesty that he reported it.

5. Free play. Because Schmidt's theoretical orientation leads him to try to provide conditions that will allow the subjects to feel at ease while at the same time motivating them to do well on the test runs, he very often allows the subjects to have free-play time on the equipment, with feedback about their hit rates. Then, should the subject feel that he or she is in a mood to do well, the subject is allowed to begin the session. Oftentimes as well, the subject is allowed to end the session at his or her whim. Were it the case that a random generator produced biased strings of targets over short periods of time, this would allow the subject, from time to time, to feel that he or she was "hot" because of an increased hit rate; the subject might then well ask to begin a test session. Once his or her score started to decline again, then the subject might be expected to ask to stop, when that was an option (e.g., Schmidt [1973]: the subject's momentary efficiency was frequently rechecked in warm-up runs before they were allowed to contribute to a test session; Schmidt [1979b]: subjects first made one or two unrecorded trial runs; if they still felt good about the test, they were then allowed to contribute to the test runs. When a subject returned for another session, Schmidt would always begin with one or two warm-up runs after which it was decided whether test runs should be undertaken or not).

6. In one experiment, some subjects chose to work as individuals, while others worked in pairs. This odd arrangement was merely reported and no reason for it was given.

7. In some cases, there was inadequate security, in that subjects were left to do test runs without the experimenter about.

8. In some cases, as Hansel (1980) has emphasized, crucial data were not recorded in nonresettable counters. In some cases, the key information is read from counters at the end of each session, instead of being automatically summarized by machine.

Thus, there are many problems that plague most of Schmidt's research, and the greatest of these is the randomization problem. However, Palmer (1985c, pp. 104–105) responded to critics' concerns about randomization tests in this way.

> The critics are correct in pointing out that Schmidt's early randomization tests do not adequately exclude the possibility of short-term biases, at least those that might occur just after the REG is activated for a run. However, the argument is weakened by the fact that the critics have so far not been able to articulate a mechanism that would produce such a bias. Short-term biases that would occur intermittently at other times would have to be consistent in direction to account for the results Schmidt found in his experiments, yet in that case they also would accumulate and thus be revealed in the randomness tests Schmidt did undertake.

What Palmer overlooks is the fact that through free play with feedback combined with optional start/stop, the subject is in a position to exploit whatever short-term biases exist, and whatever their direction. Thus, although in some studies Schmidt switched the outputs of his REG from time to time as an attempt to control for generator bias through counterbalancing, this procedure does nothing to prevent such exploitation of bias.

Evaluation

Helmut Schmidt is a highly imaginative researcher who deserves credit for his creative attempts to unravel the properties of PK. However, in my view, he has failed to demonstrate that PK or ESP exist, and without such a demonstration, all his work has been for naught, since one can hardly be successful in determining the properties of a phenomenon if one cannot demonstrate that the phenomenon exists.

My review of this database leads me to conclude that there is no evidence in any of these REG studies of any effect that needs explanation by reference to psi forces. None of the studies as they stand would be accepted for publication in a good psychology research journal, in my view, quite apart from their subject matter. They are all flawed, some terribly so. Schmidt, having become gradually more methodologically astute, should make every effort to improve his methods even more. Most likely, in my opinion, such would lead to the elimination of any significant departures from chance expectation, but that remains to be seen.

So long as Schmidt believes that feedback is important to the functioning of psi, and so long as he believes that optional starting and stopping is important to motivate the subject and to exploit his powers when he is "hot" (in order to ensure that the subject is in an appropriate mood [Schmidt 1969b]), then there will always be the danger that subjects are unknowingly exploiting short-term biases in the random target series. As long as efforts are not made to better ensure that the REG output is free of bias, as could be done using the Hansel procedure, and so long as efforts are not directed at carefully analyzing the actual target sequence, not just for one subject but across subjects and across experiments using the same REG, to discover patterns or other biases, critics will be very uneasy about accepting any explanation other than generator bias as the cause of the results. The Schmidt et al. (1986) study, while not perfect, provides a starting point for Schmidt and his colleagues to collect new data using procedures where the inadequacy of the REG is not an issue.

The Jahn Research

A much more elegant and sophisticated research program involving random-event generators has been under way for a number of years at Princeton University under the aegis of the dean emeritus of engineering, Robert Jahn, and with the participation of two psychologists, Roger Nelson and Brenda Dunne. This present review is limited to two unpublished research papers (Nelson, Dunne, and Jahn 1984; Jahn, Nelson, and Dunne 1985) that formed the basis for Palmer's (1985c) review.

Rather than employing a radioactive source, as Schmidt has done, as the basis for true randomness, Jahn uses the output of an electronic noise circuit. The noise output is filtered and amplified, and then it is sampled every five microseconds. Depending on whether the noise is above or below the zero level at that point leads to the generation of a positive or negative output pulse. In order to ensure that any residual bias is eliminated, the relationship of the sign of the output pulse relative to the sign of the noise is alternated on successive trials (or "samples," in the terminology of the Jahn team)—in other words, a single binary digit is produced.

(In keeping with the more conventional usage within parapsychology, and following Palmer's [1985] example, I shall translate Jahn's terminology into that used by Schmidt, so that Jahn's "sample" becomes "trial," his "trial" becomes "run," and his "run" becomes "block.")

In the formal test series, generation rates of either 100 or 1,000 per second are used, and each run comprises 200 trials. The count data are permanently recorded on a strip printer as well as being entered on-line into computer memory. Also, the subject receives immediate feedback via electronic displays that shows the number of runs, the number of hits in the last run, and the average number of hits since some predetermined starting point. The REG and the on-line VAX computer independently calculate the mean of each run, and the VAX also computes the standard deviation for every block of 50 runs.

The equipment can be run in one of two modes, either manual or automatic. In the former case, the machine will generate a run only when a switch is pressed, while in the automatic mode, once started, the machine will automatically initiate a block of 50 runs.

There are two types of procedure, either "volitional mode," in which case the subject chooses whether to aim for a high score (PK^+) or a low score (PK^-) in a given run, or "instructed mode," where some kind of random process determines which way the subject is to aim. There are also baseline runs interspersed ("in some reasonable fashion," the nature of which is unspecified) with the PK runs; in this case the subject is to exert no influence, so that these will serve as a randomization check. The choice of volitional/instructed mode and automatic/manual mode are "normally left to the preference of the operators (subjects), but they are encouraged to undertake additional series employing the other modes for comparison" (Nelson, Dunne, and Jahn 1984, p. 10).

The formal database consists of data from 61 series carried out on two different machines by twenty-two different subjects over a period of five years. This produced 113,890,000 trials (i.e., binary digits). These data were analyzed primarily by calculating simple t-tests using, of course, an empirically determined sample variance and comparing the observed mean to the theoretical mean. The major analysis is confined to the 390,200 runs that consisted of 200 trials per run. Thus, the mean of the theoretical distribution is 100. Ignoring the baseline runs, half the runs in this analysis were PK^+ and the other half PK^-. The mean number of hits on the PK^+ runs was 100.043, significantly greater than the theoretical mean of 100 ($p = .004$), while that for the PK^- runs was 99.965 ($p = .016$). Taken together, these two types of runs yielded a mean absolute deviation that was significant at the $p = 3 \times 10^{-4}$ level. The baseline runs (of which there were slightly fewer than of the other two kinds) produced a mean of 100.005, which did not differ significantly from the theoretical mean.

Palmer (1985c) has translated these figures into a more conventional "hit" rate by treating a miss in a PK^- run as a hit (since it is in line with the subject's goal), to yield a hit rate of 50.02 percent, which he points out is lower than the 50.53 percent mean hit rate he calculated for the Schmidt studies.

It is somewhat of an enigma for the researchers to find that the results for the baseline runs are "too good," that is, the resulting distribution of t-scores is "notably devoid of significantly high or low values, and has therefore a standard deviation well below the theoretical value" (Nelson et al. 1984, p. 25). It is conjectured that this may be the consequence of the subjects' intentions to "achieve a baseline" in the baseline condition.

Nelson et al. (1984) analyzed their data in terms of three variables: (1) manual versus automatic mode, (2) volitional versus instructed target choice, and (3) 100 versus 1,000 targets a second. They found that the significance described above was due only to the volitional runs. (This is interesting, for it suggests that subjects tend to be ineffective when the machine chooses the goal.) Results were significant regardless of generation rate, and manual versus automatic mode had no effect.

Jahn's team makes much of the individual differences in cumulative run score graphs, and they talk of a subject's "signature" that seems to identify a given subject for a given set of test parameters, but may vary for a given subject as the parameters are varied. However, these signatures are based on subjective interpretation and have not been subjected to statistical analysis. The signatures on the PK^+ and the PK^- tasks for any given subject are rarely symmetrical and often not even similar.

It is noteworthy that one subject—Operator 010—contributed 14 of the 61 series in the formal database and, again, as Palmer (1985c) points out, when this subject's series are eliminated, the remaining series in the formal group are no longer significant; this subject's scoring rate is significantly higher than that of all the other subjects combined.

Nelson et al. (1984) also report the results of 34 exploratory series, 33 of which were contributed by only two subjects. The data from only one of the subjects was significantly different from chance expectation, while that from the other was almost at the chance level. It is of interest to note that this high-scoring subject was again none other than Operator 010.

Another set of 12 exploratory series comprising 60,000 runs was carried out using a pseudo-random-event generator (a computer algorithm) rather than the REG described above. This is of some theoretical interest for parapsychologists, for if the results of the formal series were brought about by the subjects' psychokinetic influence on the output of the noise generator, then no PK effect should be observed when a strictly determined computer algorithm is generating the positive and negative pulses. Only three subjects were involved in these series. While the data from two of the three subjects did not deviate significantly from chance, significant results were obtained across the seven series in which none other than Operator 010 was the subject.

Evaluation

Jahn's team has gone to great lengths to try to ensure that their equipment is unbiased. Internal circuits are continually monitored with regard to internal temperature, input voltage, and so on. Successive switching of the relationship between the sign of the noise and the sign of the output pulse on a trial-to-trial basis was done to provide a further safeguard against machine bias. Results were automatically recorded and analyzed. Extensive tests of the machine's output and its individual components were also carried out at times separate from the test sessions. The provision of baseline trials interspersed with test trials provided a randomization check that overcame some of the weaknesses of Schmidt's procedure.

Nonetheless, despite all this machine sophistication, I still find fault with regard to procedure. An important control condition is still missing: Does the machine when unaffected by the attempted influence of the subject produce output consistent with theoretical expectation—specifically, are the baseline data in line with such expectation, for they certainly were not in the data presented by Nelson et al. (1984). A variation of the Hansel control recommended for the Schmidt studies might be useful here: One could use a random process to decide whether the next run will count as an experimental run or a control run. During control runs, the subject could be seated at the console, but doing nothing, and of course the subject would be blind as to the nature of the control run (BL [baseline], PK^+, PK^-). By accumulating scores for all three conditions, one could truly, and on a dynamic basis, evaluate the unbiasedness of the hardware. Since the subject would not know whether an experimental or control run were coming next, this would make tampering with the machine output very difficult.

Palmer (1985c) draws attention to the fact that there is no documentation regarding measures to prevent data tampering by subjects, and this is of some considerable importance since the subject was left alone in the room during the formal sessions, along with the REG and recording equipment, it would appear. It is rather uncanny that only one subject (Operator 010) accounts for virtually all of the significance in the three sets of studies. (One other subject in the formal series also produced significant results, but when Operator 010's results are removed, as mentioned earlier, there is no significant departure from chance across the total of remaining series.) I am not trying to suggest that this subject cheated; I am only pointing out that it would appear that such a possibility is not ruled out. Had the subject been monitored at all times, such a worry could have been avoided or at least reduced.

It concerns me that there is no clear indication of how the number of baseline runs was decided, and how these were interspersed when the subject was in volitional mode. Fewer baseline trials were run, relative to the other two types, across the whole database. How can we be sure that a clever subject was not able, after seeing the score for a run, to switch the designation of a high baseline run to PK^+, and the designation of a low baseline run to PK^-, thus generating significant results for the latter two types and producing baseline data that is "too good" by virtue of being devoid of the highs and lows (just as was observed in the data)? This would be more difficult to do, presumably, if the number of each sort of run were fixed in advance, and if baseline runs were scheduled on a regular basis so that any such data-tampering would be more obvious.

Palmer (1985c) also draws attention to possible problems of data selection and optional stopping. As for the former, he points out that it is not clear whether the distinction between formal and exploratory series was made in advance, and since the latter seem to be less significant, it may be that if one examined *all* the series together, the overall result may not be significant. As for optional stopping, he points out that it seems that neither the total number of trials nor the number completed by each subject was specified in advance, although it would appear from his analysis of the data that this did not have any effect. Nonetheless, this again touches my concern raised in the preceding paragraph about the looseness of the procedure in this regard.

In conclusion, despite the sophistication of the instrumentation used by the Jahn research team, there is still good reason for concern about the adequacy of the controls. A good control condition is needed to ensure that the machine truly is unbiased; but, more important, it is essential that more attention be paid to the procedure, particularly with regard to specification in advance of the numbers of trials of each sort and their temporal relationship to one another, and with regard to the security of the apparatus and data.

There is certainly a mystery here, but based on the weaknesses in procedure mentioned above, there seems to be no good reason at this time to conclude that the mystery is paranormal in nature.

2
Remote Viewing

In 1974, in the pages of *Nature*, physicists Russell Targ and Harold Puthoff described the apparent ability of their star subject, Patrick Price, to describe remote geographical locations being visited by other people with whom he had no sensory communication, a process they referred to as "remote viewing" (Targ and Puthoff 1974). In their book *Mind-Reach* (Targ and Puthoff 1977), they provide details of the Price study and of similar studies carried out at the Stanford Research Institute (now known as SRI International) in California. They claim to have carried out one hundred successful experiments, and they emphasize that anyone can become a remote viewer. Such claims for generality and replicability are quite remarkable in the context of other research in parapsychology and therefore they merit careful analysis.

In their "main series" of trials (Targ and Puthoff 1977), a total of 39 remote-viewing trials were carried out with eight subjects. There were five groups of trials, with five to nine trials each, and one or two subjects in each group. A pool of more than 100 geographical target locations (all within about thirty minutes driving range of SRI) was assembled by someone not otherwise associated with the experiment. For each trial, twelve targets were selected at random, and then from this set of targets a single target, which was never used again, was randomly selected. (As Palmer [1985c] points out, it is not clear whether the remaining targets were put back into the pool after use, nor is the basis specified on which the random selection of the target was made.)

Next, while the subject and the "inbound experimenter," both unaware of the target location, remained at the SRI laboratory, two to four "outbound experimenters" drove to the target site and spent fifteen minutes observing it. During that precise interval, the subject tape-recorded his or her impressions of the target site and drew a sketch of what he or she "saw." Once the trial was over, the subject was given immediate feedback by being taken to the target site.

For each group, the transcripts of each subject's tape-recorded descriptions along with his or her sketches were put in random order and given to an independent judge, whose task it was to visit each of the target sites used with that group and to rank-order all the transcripts according to the degree to which they appeared to correspond to the site. The sum of the ranks assigned to the response for each target was then calculated and, using exact probability tables, the likelihood of obtaining such a sum by chance was ascertained.

The outcomes were in most cases striking, and even astounding. Four of the five groups of trials produced results that were significant at the $p < .02$ level (one-tailed) or better, but more impressive were the results of the two best subjects, Patrick Price and Hella Hammid. For 7 of his 9 targets, Price's response to the target was ranked number 1, while Hammid's responses were ranked number 1 on 5 occasions and number 2 for the other 4. The odds are less than one in 30,000 that this would occur by chance!

Another series of studies, referred to as the "technology series," was undertaken to determine how much detail can be discerned by remote-viewing. Twelve trials were carried out using as targets seven different pieces of equipment housed within the SRI complex. Five different subjects participated, all but one of whom had taken part in the main series. This series was conducted in the same manner as the main series, except that targets were sampled with replacement, and only the subjects' sketches were used in the judging. Multiple drawings for the same target by different subjects were stapled together, and the judge's task was to rank each response packet against each of the seven targets. (One must wonder why only the sketches were used and why they were lumped together.) Analysis of the data revealed a remote viewing effect significant only at the $p < .05$ level (one-tailed).

Targ and Puthoff (1977) described another study, in which a subject, Hella Hammid again, was required to describe the remote targets twenty minutes *before* the outbound experimenter was due to arrive there. Although this "time travel" was judged to be a "striking success" (the formal judging procedure yielded a result significant at the $p < .05$ level), not enough methodological detail was provided to allow careful evaluation.

Critiques

The major criticisms of these remote-viewing studies can be grouped into five categories: direct cuing of the judges, nonindependence of trials, selection of data, failure to provide adequate control conditions, and subjective validation. We shall examine each of these in turn.

1. *Direct cuing of the judges:* The most extensive critical investigation of the Targ-Puthoff studies was conducted by David Marks and Richard Kammann. First, they attempted their own replication with five subjects, but were unable to produce any significant results. They had found it necessary in their research to edit out of the transcripts extraneous information that might have provided cues to the judges about which target site was visited, while Targ and Puthoff (1977) had stated that all of the subjects' descriptions were given to the judge in an *unedited* form. If there were cues that gave information about the order of the transcripts in the series, and if the judges were not given the list of target sites in a randomized order, there would be no difficulty matching up transcripts and targets, most likely without even being aware of the importance of the cues.

Targ and Puthoff (1977) had stated that all of the subjects' descriptions were given to the judges in a random order. However, when Marks visited SRI, he learned from Arthur Hastings, who had been involved in the judging of the transcripts, that the judges were provided with a list of targets arranged *in the sequence in which they had been used in the experimentation*. Furthermore, when shown the transcripts for the Price trials that had been given to the judges, Marks found that they contained a large number of cues that gave clear indications of the position of the transcripts in the series. (Example: In the transcript for the third target, a reference is made to "yesterday's two targets.")

In order to evaluate the importance of such cues, Marks rank-ordered a subset of five of the nine transcripts from the Price series against the corresponding five target locations. (Only five were used because details of the other four had been published and Marks was already aware of the pairings.) Using the extraneous cues in the transcripts, Marks was able to match *perfectly* each of the five transcripts with the corresponding target without even visiting the sites.

Subsequently, Marks and Kammann had eight experimentally naive judges repeat this blind matching for the whole series of nine targets, giving them the list of targets in the correct sequence and the randomly ordered unedited transcripts. All judges matched the first four transcripts perfectly, and for the whole set of nine, the ranking was much better than one would expect by chance ($p < .0005$).

Next, the five unpublished transcripts, edited to remove extraneous cues, were given to two judges who then visited the target sites in a random order and independently ranked the transcripts at each location. Their rankings were not significantly different from chance expectation.

As for the Hammid trials, Marks was unable to examine a complete set of transcripts, but Hastings did show him six of the set of nine, and again Hastings recalled that he had received the list of target sites in the order in which the experiments were conducted. Of the six transcripts, four *were dated*, making their relative positions in the series obvious. Other cues in the Hammid transcripts were as informative as those in the Price series. However, Hastings stated that he himself had randomized the target list after receiving it in the attempt to eliminate bias, and this would seem to have eliminated the direct cuing problem, a point emphasized by Puthoff and Targ (1981) and Morris (1980). However, as we shall see later, Marks and Kammann found fault with these trials on other grounds.

As for the other trials in the main series, Marks and Kammann were unable to gain access to the transcripts and other pertinent information that would allow a proper evaluation. They argue that it is only reasonable to assume that the same sorts of errors were also made in those trials.

Tart, Puthoff, and Targ (1980) responded to the Marks and Kammann direct cuing criticism by having Charles Tart, who was not involved in the original studies, edit the transcripts of the Price series in order to remove any possible cues, and then having the series rejudged by a new judge, presenting both the target sites and the edited transcripts in a random order. Again, seven of the nine transcripts were correctly matched, and the results were significant at the $p < 10^{-4}$ level. These authors also argued that the qualitative aspects of the descriptions, apparent direct hits in some cases, was being ignored in the focus on the statistical evaluation.

Marks (1981a) responded by criticizing the fact that the editing of the transcripts was carried out by one of the investigators in the rejudging (Tart) rather than by a neutral party, and arguing that since materials regarding four of the nine trials had been published, it is only valid to rejudge the remaining five trials.

Morris (1980) viewed the rejudging of the Price series by Marks and Kammann as an inadequate test of the remote viewing hypothesis, since the power of the test was considerably reduced by using only slightly more than half the data, an argument repeated by Palmer (1985c). Morris (1980) made several other criticisms of Marks and Kammann's critique, the bulk of which were withdrawn in his response (Morris, 1981) to Marks's (1981a) rebuttal.

Finally, Marks and Scott (1986) reported that after three years Puthoff finally released the relevant data about the series with Price. When they compared the original transcripts against Tart's edited versions, they were shocked to find that Tart had failed to eliminate a number of potentially useful but extraneous cues about the subject's location on each trial—indeed, eight of the nine transcripts retained extraneous cues. Given that the list of subject locations and the list of target sites had already been published in their correct order in *Mind-Reach*, then these cues would permit anyone familiar with that book the opportunity to match the descriptions with the targets on the basis of those remaining cues.

2. *Nonindependence of trials:* Apart from the groups of trials carried out with Price and Hammid discussed above, the other groups of trials in the main series each involved trials by more than one subject. Although Marks and Kammann (1980) were unable to examine the transcripts for these groups, they point out that Hastings stated that each subject tended to focus on certain aspects of a target site and exclude others; for example, one subject apparently tended to describe architectural and topological features while another tended to describe the actions and behavior of the experimenter. Furthermore, transcripts included the subject's name on the top of the page. Such features of the transcripts led to another error in the statistical analysis. Consider the case where two subjects completed four trials each, and the set of eight was evaluated as a unit: If the judge was aware of which subset of sites a given subject had visited, then he or she could score well above chance by ranking each subject's four transcripts against his or her four targets, cutting the odds against accurate ranking in half, which would make the rank analysis used by Targ and Puthoff totally inappropriate.

Hyman (1977b), in one of the very first critiques of remote viewing research, pointed to another problem: The statistical analysis employed by Targ and Puthoff assumes that the trials are independent of one another. Yet, he argued, such independence is vitiated by the fact that immediately after giving his or her description, the subject was taken to the target site in order to obtain feedback about how successful he or she had been. Unfortunately, this has the effect of making the next description no longer independent of the first target site: Since target sampling in the main series was without replacement, the subject might naturally tend to avoid giving responses corresponding to targets already used. For example, if the first two targets were a municipal swimming pool and a marina, then on the third trial the subject would be likely to avoid features of swimming pools and marinas. Hyman argued that, in principle, this would give a judge sufficient information to make perfect matches at each site from the descriptions. Targ, Puthoff, and May (1979) tried to defuse this criticism in saying that the target pool was large enough and the targets overlapping enough to make this problem insignificant, and Palmer (1985c) pointed out that this criticism does not apply to the technology series where sampling was carried out with replacement.

3. *Data selection:* In analyzing the transcripts for the Hammid series, Marks and Kammann found that drawings were missing for three of the six transcripts, and they wondered if the drawings had not been deemed accurate enough for the judging process. They also found a number of references in the transcripts to trials that apparently had not been reported by Targ and Puthoff in their description of the research with Hammid.

In the technology series, one to five experiments were combined from each of five different subjects, but Marks and Kammann wondered by what criteria Targ and Puthoff decided to include one trial each from three subjects, four from another, and five from Hella Hammid. They also point to indications that not all drawings were made available to the judge in this series. Since Targ and Puthoff refer in their reports to other "demonstration-type experiments," Marks and Kammann wondered whether other visitors ever tried the remote-viewing task, and if they did, were their efforts counted only as demonstrations and not experiments? They conclude that it is likely that when visitors came to try remote viewing, and the results were good, these were "experiments," but if the results were not so good, they were labeled "demonstrations," and excluded from analysis.

4. *Lack of adequate controls and control groups:* The various problems described above were in a sense made possible because there were no control trials to provide a base rate of transcript-target correspondence when no remote viewing had occurred. This rate may be con-

siderably higher than the theoretical chance level because of various methodological artifacts, such as those discussed above. Caulkins (1980) offered an example of how such studies should have been run: The subject should submit not only a sketch corresponding to his or her "perceptions" during the agent's visit to the site, but also a control sketch generated under the guise of another remote-viewing effort when, in fact, unknown to the subject, there was no target site and no outbound experimenter. Then the judges should be required to match both sketches against a photograph of the target site taken from the agent's vantage point. By comparing the ranking of the "experimental" and control sketches across a number of target sites, one would have a much clearer picture of whether or not anything extraordinary were happening.

Not only was there no control condition, there are indications of considerable carelessness and sloppiness in the running of the study and the reporting of the data. For example, Marks and Kammann (1980) draw attention to the fact that the published photographs that Targ and Puthoff have used to bolster the impact of their remote-viewing results were taken *after the fact*; it is easy to add to the illusion of a good match by choosing to photograph the site from a vantage point that highlights those features mentioned in the transcript. Again, Hyman (1977b) expressed concern that, since the subject and the team of investigators all went to the target site and openly discussed how good the description was, there is the danger that gossip might have trickled back to potential judges. He adds that similar concerns could be raised about the security of the protocols and the recording of judges' ratings.

5. *Subjective validation:* Statistical analyses aside, it seems clear from the reports that subjects and judges alike were often struck by the correspondences between the subjects' descriptions and the target sites. Marks and Kammann (1980) studied the reactions of their own subjects and judges and found that a subject and a judge might both feel very strongly about a correspondence between a transcript and a target, *even when the judge was clearly in error and was matching the transcript to the wrong target.* They concluded that, when a subject or a judge

visits a site at the end of a trial, he or she tends to notice the matching elements and ignore those that do not match, leading to an illusory validation of the remote-viewing effect.

Kammann was told by Puthoff that remote viewing results do not always reflect what the outbound experimenter actually observes on the site, but that the observer acts as a beacon relaying information, even from parts of the target site that are not visible or not actually observed directly. This, Marks and Kammann (1980) argue, makes it all too probable that one can find a correspondence between virtually any report and any of these complex targets, and once the correspondence is focused on, the process of subjective validation will work to persuade everyone—experimenters, subjects, and judges—that there was a direct hit. Moreover, if information that does not fit is not taken into account in the evaluation, then the more a subject says, the more likely it is that a hit will be subjectively perceived (Karnes and Susman, 1979).

Summary

Given these various criticisms, there should remain little doubt that the Targ-Puthoff studies are fatally flawed and that, rather than try to save something from them by arguing whether or not a given flaw pertains to a given subset of trials, remote viewing proponents should instead design and run a proper, well-controlled experiment with an appropriate control group.

Attempts to Replicate Remote Viewing

As already discussed, Marks and Kammann were unable to replicate the remote-viewing effect despite a serious effort to do so. However, there have been a number of other replication attempts, some successful, some not. Most of these have attempted to improve upon the Targ-Puthoff procedures by eliminating one or another source of artifact.

The Dunne Studies

John Bisaha and Brenda Dunne have carried out a number of remote-viewing experiments. They conducted a replication of the Hammid precognitive remote-viewing study, with Dunne as the agent, and obtained positive results ($p < .008$, one-tailed); four of the eight trials resulted in direct hits (Dunne and Bisaha 1979). While they had improved on the Targ-Puthoff procedure by having eight different judges rank one description each against the eight targets, they unfortunately employed *unedited* transcripts, which makes one uneasy about accepting their findings. (Moreover, the percipients in this study were explicitly advised to try not to define or identify what they saw with specificity, but to stick to general impressions. This contributes to the subjective validation problem discussed earlier). As Palmer (1985c) pointed out, however, the use of photographs instead of having the judges visit the sites introduced another bias, in that both the photographs and the transcripts may bear correspondence to one another with regard to indications of weather conditions. (The authors reported that they had found no such cues in the photographs.) It would also appear that Hyman's (1977b) concern that, following feedback, subjects would avoid descriptions that might apply to the previous targets was not ruled out.

Bisaha and Dunne (1979) conducted two additional precognitive remote-viewing studies. Each produced significant results in line with the remote-viewing hypothesis. The more interesting of these involved transatlantic remote viewing. Each morning over a period of five consecutive days, the percipient, in the United States, attempted to describe where the agent, in Eastern Europe, would be 24.5 hours later. The agent was then to spend 15 minutes at the appointed time the next day attempting to concentrate on the surroundings and taking a photograph that would later be compared against the percipient's description. Upon his return the agent gave the five photographs and brief descriptions of the target sites, in random order, to the percipient to rank order in the usual way. The percipient also gave to the experimenter a set of transcripts, in random order, for the experimenter to rank against the targets. Finally, a third person, who had no other connection with the experiment, also rank-ordered the photographs against the descriptions. All three rankings were significant ($p < .025$, .025, .05, respectively, all one-tailed).

Dunne, Jahn, and Nelson (1983) set about to develop an analytical scoring technique for remote-viewing studies. Both sender and percipient were required to code their perceptions of the target in terms of 30 binary descriptors (such as indoors or outdoors). They then empirically derived a baseline distribution of chance scores by the analysis of 42,000 mismatched permutations of targets and perceptions. Next, they examined the data from some 300 separate remote-viewing trials, most of which were of a precognitive nature, which ranged over physical separations of up to 11,000 miles and time intervals of more than 48 hours. They applied several different scoring methods to assess the correspondence between two sets of 30 binary bits and found that the composite z-scores calculated for the total sample were highly significant regardless of the method of scoring. As Palmer (1985c) has already pointed out, since no procedural details about most of the trials are included in the report one cannot offer a methodological critique.

The Schlitz Studies

M. Schlitz and E. Gruber (1980) conducted a long-distance remote-viewing experiment with Schlitz, in Detroit, as percipient and Gruber, in Rome, as agent. Gruber, over a period of ten consecutive days and at preset times, visited various targets in Rome that had been selected randomly from a target pool, without replacement. At the times that Gruber was at the target sites, Schlitz recorded her impressions. Following the completion of all ten trials, Schlitz sent a copy of her ten protocols to Gruber, still in Rome, and *he and another individual who was blind to the targets translated the protocols into Italian.* They also checked for cues in the transcripts that might allow judges to infer temporal order, but none were found. The percipient's sketches were photocopied and attached to copies of the protocols.

Each of five judges rank-ordered the protocols against the target sites. Highly significant results were obtained for the combined rankings of the five judges, and also separately for four of them.

To eliminate the possiblity that the agent and percipient, being people of shared interests, may both be attuned to events going on in the world about them which will lead them, at any given time, to focus more on certain characteristics (such as the weather), thus leading to an artifactual correspondence between their reports, the authors carried out a rejudging of these data (Schlitz and Gruber 1981). Two new judges in Rome were asked to rank the percipient's impressions against the target sites, but were not given the agent's impressions. The scoring was carried out as before, and the results were found still to be significant but at a reduced level.

As Palmer (1985c) observed, it is not stated in the report whether the transcripts were given to Gruber in random order or whether the judges received the target list in random order. It is also unacceptable that the translation and editing of the transcripts involved Gruber, who knew the order of the target sites. For all these reasons, this study would not have been accepted for publication in a good psychology journal, its parapsychological nature aside.

In a subsequent study (Schlitz and Haight 1984), these problems were eliminated. Schlitz, in North Carolina, served as percipient while her coexperimenter served as agent in Florida. Ten target sites were randomly chosen without replacement from a pool (number unstated) of target sites, and the agent visited a different target site for 15 minutes on each of ten days; at these same times, the percipient recorded her impressions of the agent's location. Agent and recipient sent their materials (typed transcript of tape-recorded impressions from the percipient, and final target order from the agent) to a third party, who randomized both the transcripts and the list of locations and sent them, along with rating sheets, to two judges. There were no notes from the agent; there was no direct communication between agent and percipient until after the experiment was over, and the percipient received no feedback until well after the experiment was over. Subsequently, two judges went together to each target site and evaluated the correspondence between the site and the percipient's description. The results were significant at the .05 level, one-tailed. This study is the best controlled of all the remote-viewing reports examined so far, a point upon which I am in agreement with Palmer (1985c).

The Karnes Studies

Karnes and Susman (1979) conducted a remote-viewing experiment within the framework of a signal-detection experiment. Signal-detection theory provides a powerful method for detecting a signal, if one is present. Unlike the other studies mentioned so far in this survey, this study was set up with a control condition. Moreover, rather than depending on the vagueness of the percipient's reports and the arbitrariness of the judging, this study presented each percipient with a response booklet in which 18 sites were represented, each on a separate page, by four photographs of the target site taken from different vantage points. While the agent was on the site, the percipient was to select one or more of the 18 sites as possible locations for the sender, and to rate the confidence of each selection on a five-point rating scale. Of the 18 sites, only 9 were actually used, and the others where "noise" sites. In addition, there were two other sites that were not represented in the set of 18, and these were used to provide a sort of control condition. There were a total of 10 receivers for each of the 9 targets and 25 receivers for the two control-group targets. The results did not statistically differ from chance expectation.

Karnes, Ballou, Susman, and Swaroff (1979) conducted a remote-viewing experiment with a different sort of control, this one a full-fledged "control group." Each subject participated in two remote-viewing trials, but half of the subjects did not have any contact with the sender before the first attempt, whereas the other half of the subjects met the sender immediately prior to the commencement of the trial. The authors reasoned that if the receiver is unaware of both the identity and location of the sender, then the receiver's impressions should be self-induced and would serve as a control for guessing (response bias). After the first trial, the receiver and the sender met at the target site and there was no limitation placed on their discussion. The second trial was identical for both groups of subjects and similar to the traditional remote-viewing trial.

Judging was carried out using both a confidence rating scale and a rank-order scale. A set of 120 judges evaluated the results for trial 1, and another 120 evaluated the results for trial 2. No statistical support for remote viewing was found. In trial 1, there was no significant difference in the scoring rate of the experimental and control groups.

In another study (Karnes, Susman, Klusman, and Turcotte 1980), the subjects were eight individuals who claimed to be psychics. Again, no evidence of remote viewing was found.

Evaluation of the Remote-Viewing Studies

My own view is that Marks and his colleagues have so clearly pointed out the deficiencies in the Targ-Puthoff studies that there is no point in trying to analyze them further. Added to the clear problem of sensory cuing and the conjecture about data selection is the fact that Targ and Puthoff have been very reluctant to provide access to their raw data, and have only just this year released a portion of it after considerable pressure was brought to bear. The result was to show that the editing that Tart had supposedly done very carefully was actually done carelessly and did not eliminate the problem it was intended to solve.

Of the Schlitz studies, the Schlitz and Haight (1984) study was better controlled than their earlier ones and those of Targ and Puthoff; it is interesting to note that the results in this case are much less striking than when conditions were less well controlled. However, one can have no confidence at all that the significant results reported in this study have any real meaning, for there was no control condition to indicate the background "coincidental" rate. The same applies to the Dunne studies, quite apart from the flaws previously noted.

Karnes and his colleagues used control conditions and found no evidence of remote viewing. This does not by itself mean that the phenomenon does not exist, for one cannot prove nonexistence. However, their work clearly points to the proper methodology to be used in the study of remote viewing. One must use a control condition for establishment of a baseline "guess" rate. Without such a condition, the statistically significant results of the other studies are not interpretable, for they are based on an *assumed* guess rate—which, because of various sources of artifact already discussed, and perhaps others not yet discovered, may well underestimate the actual guess rate.

The lack of proper control conditions would keep the Targ-Puthoff, the Dunne, and the Schlitz studies from publication in a good psychology journal, the parapsychological theme quite aside. Until remote viewing is demonstrated, and replicated, in well-designed studies executed under well-controlled conditions (employing control groups), there is no reason at all to take remote viewing seriously.

3
Overall Conclusions

This examination of random-event-generator and remote-viewing studies leads me to the inescapable conclusion that none of this research has served to demonstrate the reality of psi phenomena. Instead, there are serious flaws and shortcomings that require elimination before one can have any confidence in the statistical departures from expectation.

If one had to single out the most serious and recurrent problem, it would be, in my view, the absence of proper control groups or control trials. The use of the Hansel procedure in the Schmidt studies would generate control-group data that would allow a direct comparison of the "experimental" data with the baseline, coincidental "background" rate of scoring as empirically determined in the control trials. I have suggested that a modification of this procedure could be used in the Jahn paradigm. More traditional control trials have also been suggested for the remote-viewing research.

Psychological research is built around the concept of comparison of experimental and control conditions. There is every reason to demand a similar approach in parapsychology. The arguments that psi forces cannot be turned off and may equally affect experimental and control trials leading to no difference between the two is not acceptable. In the absence of conclusive, or even persuasive, evidence that psi exists, there is no reason to suspend the very procedure that could serve as our strongest bulwark against erroneous interpretation of data. If the "phenomenon" disappears in such a case, then we should bother no more about it.

Appendices

Appendix 1

Detailed Critique of the Schmidt Studies

Schmidt (1969a): "Precognition of a quantum process"

In this report, Schmidt presented the data from two experiments that examined the abilities of subjects to predict or influence the outcome of a quantum process. The statistical evaluation of success in these experiments is based on the likelihood of obtaining a hit by chance if the target series is random. As is typical with Schmidt, he was the sole experimenter.

The subject was seated in front of a panel containing four colored lamps and four corresponding push-buttons. Prior to a button being pressed, an electronic circuit was in operation such that electrical pulses, at the rate of one million per second, arrived at an electronic four-position switch, and each pulse advanced the switch one step, in the sequence 1, 2, 3, 4, 1, 2, 3, 4, ... Once any of the four buttons was pressed, there was a short waiting time of unpredictable length, determined by the arrival and registration of an electron from a decaying strontium-90 source. At this point, an electronic gate closed so that the switch stopped at its current position and the corresponding lamp was lit. Once the switch stopped, one mechanical counter advanced by one to indicate the number of trials, while a second such counter advanced by one only if the illuminated lamp corresponded to the button that was pressed (a "hit"). These two counters were nonresettable, and their readings were recorded by hand. Furthermore, an external paper-tape punch recorded which lamp was illuminated and which button had been pressed.

First Experiment

Here, the subject's task was to try to predict which of the four lamps would next illuminate, and the aim of the experiment was to see if subjects could make such predictions at a rate significantly above what would be expected on the basis of chance alone. Schmidt points out that, although the study is set up to examine precognition, it is impossible to rule out the possibility of psychokinesis: The successful subject could be manipulating the quantum process so as to increase the likelihood that a particular light will illuminate.

There were three subjects chosen from a set of 100 potential subjects on the basis that they were among those who seemed to consistently score above-chance in preliminary trials. Rather than presetting the number of trials, Schmidt for some unspecified reason assigned a *range* of trials to each subject: one subject was to do between 15,000 and 20,000 trials, while each of the other two were to do between 20,000 and 25,000 trials. Except for some of the trials with one subject, Schmidt was present during all the trials.

Randomization checks: Five million numbers were generated by the REG for control purposes, and the frequency of all four numbers and all sixteen sequential pairs was calculated; these frequencies did not differ significantly from chance expectation. These five million numbers were recorded on 100 different days, "preferably directly after the experimental sessions."

Results: The 63,066 trials from all three subjects combined yielded a hit rate that was significantly above chance expectation ($p < 2 \times 10^{-9}$), although the actual hit rate was 0.261 versus the chance expectation of 0.250.

Evaluation: There were a number of weaknesses in this study, including the following:

A. The randomization checks were inadequate. Although Schmidt tries to reassure the reader that he carefully checked to ensure that the generator operated without bias, he did not check beyond the doublet level, and he did not check for short-term biases. We are given no information about the temporal relationship between the control runs and the tests except for the "preferably after" comment.

B. The subject was free to "play" with the equipment (with paper punch and nonresettable counters disconnected) and to decide when to start and stop a given session. If there were short-term biases in the generator that lasted, suppose, for ten minutes, and which were not detected during the randomization tests (which were of much greater length), these play sessions, the feedback, and the freedom to choose when to start and stop a session would provide magnificent opportunity to exploit, consciously or unconsciously (and most likely the latter), that bias. After all, the subject would want to begin a session, presumably, when it appears that he or she is "hot," while, if the subject's scoring rate declines, he or she may well want to end the session and start again later. The subject was given immediate feedback by means of a set of resettable counters (distinct from the nonresettable ones mentioned above) that displayed the number of trials and the number of hits.

C. Methodological sloppiness: It is sloppy and totally unnecessary to assign different numbers of trials for different subjects. It is equally so to preset a range rather than an actual number of trials. Worse, one finds upon examination of the data that the subject whose range was 15,000 to 20,000 trials actually completed 22,569 trials, while one of the subjects who was supposed to carry out 20,000 to 25,000 trials actually carried out only 16,250 trials. This sort of looseness by itself would make it difficult to obtain publication in a good psychological journal. However, given the long debates in parapsychology about the optional-stopping problem, one wonders why a parapsychological researcher would allow himself to build such optional stopping into his procedure, even if it can be shown that it would not affect the data very much. It is likely that Schmidt did so in the belief that to do otherwise would detract from allowing subjects to operate when at their best.

There was again sloppiness in the distribution of sessions, which were at the subject's whim, it appears. There were 18 sessions (11 for one subject, 5 for the second, and 2 for the third) and we are not told over how many days these sessions were spread. Moreover, the study was carried out, for no apparent reason, with very few (3) subjects, despite the pretension of having selected a "team" from 100 potential subjects.

D. Recording: As Hansel (1980) has pointed out, Schmidt used an automated machine, but then left the recording procedure vulnerable to recording errors since it was necessary to manually read and transcribe the nonresettable counters.

E. Analysis of data:

(1) In his data analysis, Schmidt pools the data from all the subjects. One of the three subjects in the first experiment did not score significantly higher than chance, but his data were put together with the others to yield overall significance. The problem with this approach is that if one subject, for whatever reason, were to score very highly—and this might in some instances be because of methodological artifact or even fraud—then the pooling of data might yield overall significance, whereas it might be more reasonable simply to point out that one subject scored remarkably high, while others did not. My objections here are perhaps picayune, but I must ask this question: Why use a number of subjects if indeed they are interchangeable? Why not do the whole study with one subject? Indeed, as will be seen later, on occasion Schmidt does exactly that.

(2) No information is provided about the proportions of singlets and doublets in the actual target series, and given what is being claimed, this is a serious lacuna. It would illuminate the discussion of possible short-term biases in the REG to know whether or not the target sequences deviated very much from chance expectation. Schmidt does point out that one subject in the first experiment (who reported having attempted to influence the outcome rather than just predict it) actually experienced a target sequence in which the red light (corresponding to the number "4") came on significantly more frequently than would be expected by chance. This finding is damning evidence with regard to the inadequacy of the randomness checks (unless one begs the question and assumes the existence of the phenomenon that Schmidt is trying to demonstrate,

in which case, as Schmidt does, one could argue that this nonrandomness is due to, and evidence of, PK). If we do not assume a priori the reality of psi, then we must conclude that at least for the highest scoring subject, the sequence of targets was nonrandom. The subject's report that, on his own and without instruction, he concentrated on causing the red lamp to come on is quite consistent with the idea that he was differentially rewarded (with a hit) when he chose red. One would like to examine the data for the other subjects as well in order to explore whether or not such departures from randomness were evident in the target series presented to them as well.

One must wonder what interpretation Schmidt would have given to the excess of reds had the subject not mentioned that he had tried to use PK, or if the red light had lit up significantly *less* frequently than expected; one must also wonder whether or not the subject made mention of PK before learning that one light had come on significantly more frequently? Note that the same subject was the highest scoring subject in the first experiment and carried out all his trials in only two sessions.

Second Experiment

Here, subjects were allowed to choose to try for a high score (high number of hits) or a low score, and we are told that "at the beginning of each session it was decided whether to try for a high score or a low score" (p. 107). However, of the three subjects, two of whom had taken part in the first experiment, only one apparently ever chose to go in more than one direction. Of the remaining two subjects, one always went high and the other always went low. These latter two subjects contributed 5,000 trials each while the former contributed 10,000, about 57 percent high and 43 percent low.

Results: Again combining all the data, Schmidt found an overall scoring rate of 27 percent, which is significant at the $p < 10^{-10}$ level.

Evaluation: The same criticisms about randomization checks, small number of subjects, and so on, that were made in the analysis of the first experiment also apply here. As well, allowing the subject who went high on some trials and low on other trials to make that choice, presumably following the "play" period, again makes tuning in to short-term biases more likely. For example, if red seems to occur more often than it should, then one might aim high and choose red a lot; if red seems to be relatively infrequent, one might then aim low and choose red a lot. It would be very interesting to look at the raw data and as a first measure examine the distribution of subjects' choices as well as the distribution of targets. These data were not analyzed in the report.

Another weakness, as Hansel (1980) has pointed out, is that recording errors were made quite possible by the fact that high and low scores were not separately recorded in the nonresettable counters. Indeed, there was *no* overall deviation from chance if one simply examined the totals shown by the nonresettable counters, and the alleged psi effect was only apparent when the scores were broken down on the basis of whether or not the goal was high or low scoring. Hansel was very concerned that so much manual data analysis intervened between the results shown on the nonresettable counters, which indicated no overall deviation, and the significant departures from chance that were evident following the manual data assignment and analysis.

Overall Judgment

The weaknesses discussed above reduce these two studies to the status of, at best, pilot studies; the criticisms need to be addressed before there is any reason at all to take the psi claim seriously. This report, in my judgment, were it merely dealing with normal psychological phenomena, would not be accepted for publication by peer-reviewed psychology journals simply because of the sloppiness of the design and execution and the lack of thorough randomness checks. With regard to the latter, and as mentioned above, one would need randomness checks that are both of the same length as the target runs and run during approximately the same time periods as the experimental runs.

Schmidt (1969b): "Clairvoyance tests with a machine"

Schmidt, again the sole experimenter in this research, views this experiment as a continuation of the 1969a studies. However, rather than using a "real-time" random generator that allows for the operation of PK,

Detailed Critique of the Schmidt Studies 137

Schmidt used as his target sequence a set of 100,000 digits (1,2,3,4) taken from the Rand tables and then further shuffled and again checked for randomness. These numbers were punched into paper tape. Schmidt argues that this rules out PK, allowing only precognition or clairvoyance to operate. (It is interesting to note that Schmidt obviously denies PK the power to alter the pattern of holes in a paper tape.)

Six subjects, two of whom had participated in the 1969a experiments, participated in this experiment. Although there is no explanation as to why this was done, four of these subjects chose to work in two pairs, so that one subject would do a number of trials until he or she felt like stopping, then the second subject would work at it until he or she wanted to stop, and then the first would resume. This would be repeated until they decided to end the session. A subject chose to aim for hits or misses on any given run and then a switch was thrown in order to record the run on paper tape. As Hansel has pointed out, again the high-low distinction was not recorded in the nonresettable counters.

There were no limits on the number of trials to be contributed by any given subject, or on the length of the sessions. It was decided in advance to run either 15,000 or 30,000 trials, although no reason is given as to why these two figures were specified and 15,000 were actually run.

Randomization checks: The randomness check is not described in the report, but a reference is made to an unpublished paper.

Results: There were +108 deviations on the high trials and −152 on the low trials, for an overall success rate of 26.7 percent versus the chance rate of 25 percent. (This is statistically significant, $p < 0.6 \times 10^{-6}$.)

Evaluation: Again, there is a lot of sloppiness here: Some subjects worked in pairs, some did not; subjects chose when to begin, end, or switch, and there was no regulation of the number of trials per subject, and so on. However, the key question is this: If one were to examine the actual sequences of targets, were those sequences (which were short, relative to the overall set of 100,000 numbers that had passed Schmidt's randomness checks) unbiased? Using Schmidt's own words, PK is not a factor here, and so if the overall frequencies associated with singlets, doublets, and even triplets are not as one would expect from a random series, then again we might expect the subjects to learn to exploit the biases in the series.

Overall Judgment

Given the lack of information about the randomness checks, and given that the actual target sequence was carefully analyzed after the fact, the conclusions drawn by Schmidt are premature, and this paper would be unlikely to be accepted for publication in a peer-reviewed psychology journal even if only normal psychological processes were being investigated.

Schmidt (1970a): "A PK test with electronic equipment"

With this study, Schmidt, again the sole experimenter, switched from the four-lamp modulus-4 random generator to a nine-lamp binary generator. The generator produced a random sequence of +1s and -1s, based on radioactive decay. The subject viewed a circular display of nine lamps, only one of which was illuminated at any given time. The random event (+1 or -1) determined the direction of progression of successive illuminations. The subject's task was to choose a direction and then try to make the lamps advance in that direction. If the subject chose counterclockwise, a switch was flipped to make the lamps go that way on a +1, but a +1 still registered as a +1.

Randomization checks: Again, as in the earlier studies, success is demonstrated by the degree to which the correspondence between the subject's goal and the outcome exceeds what would be expected by chance. Thus it is of utmost importance to demonstrate that the random-number generator is free of bias, and to this end, Schmidt generated 4 million numbers on many different days (the relationship of these days to experimental sessions was not specified). The number of +1s, -1s, and "flips" (i.e., a change from +1 to -1 or vice versa) was then examined, and no bias was detected. Schmidt, presumably showing sensitivity to criticism of his earlier experiments with regard to his having used control sequences that were much, much longer than the target sequences, also divided the numbers into 10,000 sequences of 400 and looked at the number of +1s, -1s, and flips in each of them. The outcome of this

Detailed Critique of the Schmidt Studies 139

was consistent with the normal distribution. (However, it is not so much the overall distribution that is important here, but rather the *sequences*: Do a number of high 400 sequences occur in a row, for instance?)

An additional measure was taken to protect against bias: After the first half of the confirmatory test, the two outputs of the generator were interchanged internally, so that any bias in favor of one digit would be reversed. (We are not told what the breakdown of scores was before and after this change. Could it be, for example, that the deviation was *highly* negative pre-change, and only barely positive post-change, thus giving an overall negative score, but one much less impressive than in earlier studies [1969a, b]?)

Results: Although Schmidt ran some preliminary trials with 18 subjects, and reported and analyzed the data, such trials should be viewed only as pilot tests and not of concern to us here. In his "confirmatory series," fifteen subjects took part. The number of trials was preset: There were 64 sessions of 4 runs each, with each run consisting of 128 binary numbers. Subjects, as is typical in Schmidt's experiments, contributed varying amounts of data. No individual results were reported. Overall, there was a *negative deviation* of 302 hits, which was statistically significant ($p < .001$, two-tailed). However, this is precisely what Schmidt expected! Negative scoring was indeed the norm in the preliminary trials, and this led Schmidt to predict negative scoring and to motivate the subjects to score negatively and to only use subjects who did score negatively in the preliminary trials (all but one of those subjects did just that).

Schmidt reversed the outputs of the generator halfway through and stated that if there was a systematic bias, this would compensate. However, no data are provided for the target sequence before and after this change. Also, while we are told that the randomization checks rule out biases, short-term biases are not ruled out by this procedure. Schmidt almost alludes to this himself when he discusses the possibility that the study might be tapping precognition (of the subject *or* the experimenter) rather than PK:

> Since the sequence of generated numbers depended critically on the time when the test run began, and since the experimenter, in consensus with the subject, decided when to flip the start switch, precognition

might have prompted experimenter and subject to start the run at a time which favored scoring in a certain direction. (p. 181)

Evaluation: Again, it is disappointing to see Schmidt change procedures without really having done more than a cursory investigation with the earlier ones. Here Schmidt switched to a *binary* generator (after having used his modulo-4 twice, and the Rand tables once). As well, because only one subject, in the preliminary sessions, really scored high in a positive direction, it was decided that, for some reason, negative scoring was "in," and so negative scoring was predicted for the "confirmatory" trials. The high scorer was eliminated and negative thinking was encouraged. The subjects who scored most negatively were used most in the confirmatory series, and new subjects were added only if preliminary testing suggested negative tendencies in scores. Why? How can the task suddenly become a negative one? Or is this all because that initial group of subjects just happened to be largely negative?

The subjects were encouraged to think pessimistically and in terms of failure. Yet Schmidt alluded in this paper to the notion that PK is goal-oriented—even in a complicated set of circumstances, results are obtained by concentrating only on the goal. Here the goal is self-contradictory: Subjects were supposed to try to influence the lamp to go in the direction of their choice, but they were also supposed to want to fail! Why not have them concentrate on having the sequence go *opposite* to their preferred direction? It is actually worse than that; if the subject chose to try to make the lamps light in a counterclockwise direction, a switch was flipped to cause a +1 number to move the illumination of the lamps in counterclockwise direction, so that failure (an excess of –1s), which is *really* success (because subjects are encouraged to psi-miss), is now linked with perceived success on the board, whereas when the subject chose clockwise, failure (excess of –1s, again which is really success) is associated with perceived failure on the board. What is the goal-directed PK going to do?

An effort was made this time to prespecify the number of sessions and trials; however, the number of sessions still varied between subjects. This is not a fundamental problem, but it indicates a certain sloppiness that is unwelcome in experiments that are supposed to demonstrate the existence of a phenomenon that seems to defy the contemporary scientific worldview.

Overall Judgment

Again, much more work is needed to explore this situation to see just what exactly is going on. Schmidt accepts his findings as evidence of psi and then simply moves on to something else. In my judgment, this paper, quite apart from its parapsychological nature, would not be publishable in any good psychological journal.

Schmidt (1970b): "A quantum mechanical random-number generator for psi tests"

No experiments are reported here. This paper presents a description of Schmidt's binary RNG and a discussion of his randomness checks.

Schmidt (1970c): "PK experiments with animals as subjects"

This report consists of two studies, one carried out with Schmidt's pet cat and the other with a number of cockroaches. In the first case, the cat was placed in a cold garden shed, and a lamp, which when illuminated provided some warmth, was either turned on or off, depending on the output of the binary random generator. There were 1,000 trials in each of ten sessions. In the second case, cockroaches were placed on a grid through which shock was or was not delivered, depending again on the output of the binary generator. There was an exploratory series of 25 sessions, followed by a confirmatory series of 100 sessions, with 4 runs per session and 64 numbers per run, all this preset.

Randomization checks: In the series with the cat, these consisted of running the RNG for 8 hours per night for 24 nights, with the lamp *outside* and the complete system left running. No bias was evident in these 691,200 trials. Between sessions, we are told, the RNG was left running continuously to verify the lack of bias. In the cockroach study, the RNG was left running each night after the PK test; again no bias was found. As additional protection against bias, *both* generator outputs used equal number of times to produce +1 (shock). (The report does not explain whether the reversal was made halfway through the experiment, or at various times.)

Results: In the study with the cat, the first five sessions yielded an above-average rate of +1s (lamp on) (CR [critical ratio] = 2.42). However, the next five sessions yielded results that were (insignificantly) below expectation. In the cockroach study, it was found in the confirmatory series, as in the exploratory series, that the cockroaches received significantly *more* shocks than would be expected by chance ($p < 10^{-4}$).

Evaluation: These were very poorly designed and executed studies. Again, there were no short-term randomness checks. (Test sessions were only a half-hour in length.) It was not specified (or at least not stated) in advance that there would be ten trials with the cat. On what basis were the first five separately analyzed and why was there no overall analysis of the ten trials taken together? Indeed, one cannot take the results of the first five trials, an arbitrarily chosen block, to be indicative of anything, especially given the results of the next five trials.

Schmidt readily generates ad hoc explanations: During the second five trials (for which the data were *not* significant), he comments that the cat seemed disinterested in the lamp. Would this same observation have been forthcoming had this subset of trials been significant? Further, Schmidt suggested that the reason the cockroaches received more shocks than expected by chance may have been that it was *his* PK and not the insects' that was the psychic influence, and given his dislike for cockroaches, this led to excess shocks. However, he alluded to another experiment with cockroaches, which ran automatically without the experimenter present and produced similar results. He stated that these would be subsequently published, but to my knowledge, they have not been.

Overall Judgment

This study is clearly inadequate as a demonstration of anything having to do with psi. This paper would never have been accepted for publication in a mainstream psychology journal because of the methodological weaknesses and general carelessness. Had Schmidt thought the results to be as compelling as he suggests, one would expect to see a series of further studies of the same sort. That is not the case.

Schmidt and Pantas (1972): "Psi tests with internally different machines"

This study was designed to try to differentiate between precognition and psychokinesis in a psi task. Subjects used a test machine that involved choosing which one of four lights would be next selected by the random-event generator. In the precognition mode, the machine worked just as in the 1969 studies; pressing one of the four buttons activated the random-event generator and one of the four lights subsequently was lit. In the PK mode, the light corresponding to the depressed button lit up, indicating a hit, only if the internal generator produced the number 4 (from among its range of 1, 2, 3, and 4). Thus only by psychokinetically forcing the production of 4s could the subject increase the hit rate. (Schmidt admits that these are not *pure* precognition or PK modes, for PK could be involved in the precognition mode to influence the generation of a number corresponding to the button depressed, and in the PK mode, one could use precognition to wait until the chances for obtaining the desired target were good.) The mode could be changed by a flip of a switch, but the subject would be unaware of any change, since the task as presented to the subject was identical for the two modes.

First Experiment

In the first experiment, the goal was not psi-hitting, but rather psi-missing. Schmidt writes:

> Psi-missing had been observed in preliminary tests in which the subject had to perform in front of a group that reacted with friendly laughter to each of the subject's misses. (p. 226)

One might well wonder what would motivate Schmidt to create such a test situation: Schmidt tells us that the subject was instructed to try to *miss* rather than hit, and so the goal was psi-missing. Thus, the negative influence of the laughter upon the subject's misses (which are, in this case, successes) then pushes the subject toward more hits (which in this case are failures, or misses). Thus, the interpretation given to the above-chance hit-rate is that the subject whose goal was psi-missing actually "psi-missed" that goal, producing psi-hitting! It is not clear just exactly how the decision to study psi-missing in this experiment came about. (Were the three pilot groups actually asked to avoid hits? It seems so, since Schmidt says that the usefulness of this procedure was *suggested* by their results, and since they scored above chance then it would seem that they had not been instructed to try to score above chance. But why would he ask people to psi miss in the first place?)

Schmidt ignores the logical morass created by his two-mode precognition-PK machine. In the precognition mode, all is well: If precognition exists, the subject foresees what lamp will light next and then presses the corresponding button. Alternatively, if PK is used, the subjects could supposedly arbitrarily choose a button and then exert psychic pressure to cause the corresponding lamp to light. In the PK mode, however, things are different. If the subject uses only PK, then he or she must learn that success occurs only when 4s are generated, and then the subject must bend his/her psychic influence to produce more 4s. Consider, however, what would happen if a subject tried to use precognition in the PK task. (Note that the reason Schmidt calls this a PK mode is because it would be difficult to succeed merely by precognition.) It is more than merely difficult: It leads to such logical confusion as to put the possibility of precognition into question. If precognition involves seeing what lamp will light next, and if one presses the button beneath that lamp, then if another lamp lights because the random generator did not produce a 4, one did not foresee the lighting of the right lamp. Suppose precognition were perfect: one would know then which lamp would light up, but if one presses the button for that light, unless one is somehow constrained to wait for a 4, one would be defeating the precognition.

Schmidt was the primary experimenter in the first experiment, but in most tests, we are told, a second experimenter (one of five people) was also present. Subjects consisted of 18 groups of students, teachers, etc., whose participation usually followed a lecture on psi; the testing was done outside the laboratory (presumably at the place where the lecture was given); although we are told in the introduction that the machine allows switching back and forth between modes, "for practical reasons" each group worked in only one mode. The mode was switched for each subsequent group. Only the principal experimenter, Schmidt, knew that the experiment included two types of tests. There were a total of 214 subjects in the precognition group and 157 in the PK group; there were varying numbers of subjects per group, and data is presented for groups, not individuals. It is an example of the carelessness of the preparation of the report that one only discovers when one reads a footnote to the table of results that each individual participated only until he or she obtained a hit. Only 500 trials were planned for each mode, but 740 trials were run in the precognition mode (another disquieting indicator of methodological sloppiness), and so the last 240 were not used in the analysis. (My check shows this does not work in favor of Schmidt's hypothesis; his results would be stronger if he had left the 240 additional trials in.)

Randomization checks: None described.

Results: The scoring rate in the precognition mode was 29.8 percent, significantly higher than the 25 percent chance level ($p < .01$, one-tailed). In the PK mode, the scoring rate was 31.4 percent, again significantly higher than the 25 percent chance rate ($p < .0005$, one-tailed).

Second Experiment

Schmidt was the only experimenter in the second experiment, which involved only one subject, coauthor Pantas. After a pilot test 500 precognition trials were run followed by 500 PK trials; these were run in blocks of 25 per session; usually 1 session per day. Pantas first was tested in a precognition pilot test (350 trials), and then a "confirmation" precognition series (500 trials), and finally a "confirmation" PK series (500 trials). Pantas was left alone with the machine for 20 minutes to practice Zen before each session. Schmidt was not present in the room, but by monitoring the paper punch "could follow the progress of the test whenever he wished without disturbing the subject" (p. 231).

Randomization checks: None described.

Results: Pantas scored at about the same above-chance rate in all three series, including the pilot (pilot, 30.9 percent, significant at $p <$ .01; precognition, 32.8 percent, $p <$ .00005; PK, 30 percent, $p <$ .005, all levels one-tailed).

Evaluation: No mention was made of any randomness checks in these two studies, but one would presume that Schmidt must have carried out the same kind of inadequate control runs that he carried out for earlier studies. Obviously, the above-chance scoring rates are only of interest if one can be certain that there were no biases in the target series. Again, the Hansel control procedure (i.e., taking target runs two at a time and randomly assigning one to the experimental series and the other to the control series) would be invaluable. It would be of considerable interest to examine the actual target series used in the various tests. For example, if PK is actually being used to produce more 4s in the PK series, one might expect that the PK target series would include more than 25 percent 4s (as was apparently the case, given the approximately 30 percent hit rate), while one might expect that this would not occur in the precognition series, where there would be no need for it. On the other hand, if one found a similar excess of 4s in the precognition series, this might well tempt one to disbelieve that PK had been at work in the PK series. Indeed, it is interesting that Schmidt arbitrarily chose 4 as the number to force in the PK tests, for one of his subjects in an earlier study had reported that he had tried to generate more 4s (going against the instruction to try to predict which lamp would next light up), and Schmidt found more 4s in his target series. The skeptic might suggest that Schmidt's machine is prone to a short-term bias that boosts the production of 4s. One would like to examine the subjects' responses in the precognition series, for if they obtained their above-average score rate through "response-matching," one would expect that they would have learned to depress 4 more often than the other buttons, and one should see this in their responses. On the other hand, the PK set-up did not allow for response-matching, and the subject could hit any button and still obtain an above-chance score of 30 percent as long as about 30 percent of the target numbers are 4s.

Again, there were no controls; one would like to see, for example, half the subjects attempting to demonstrate psi-hitting, and the other half psi-missing. However, Schmidt might well argue, as he suggested in the introduction, that these subjects for whatever reason were more attuned to psi missing. This is unsatisfying.

Note that for the first experiment there was no paper punch; yet this is probably not why Schmidt ran the system only in one mode or the other for any given group, rather than switching back and forth without the subjects' knowledge. The paper punch probably did not record the mode, for although it was used in the second experiment, which used only one subject, all the precognitive trials were run and then all the PK trials. Does this not mean that Pantas could have changed back and forth between modes had he wanted to, without the experimenter knowing (not that this should make any difference)? At any rate, the procedure is certainly vulnerable to recording errors since the recording and computations are not automated.

Overall Judgment

Once again Schmidt has served up an empirical report where he makes a number of shifts from earlier studies: a new machine arrangement; testing carried on outside the laboratory, with no paper punch to record results—at least for the first experiment; double psi-missing as the goal (the subject, whose goal is to psi-miss, is actually treated in a way that might lead him or her to psi-miss the psi-missing, thereby yielding psi-hitting); a subject continues until he encounters a hit (i.e., really a miss) and then his participation is over. Not only are there these changes, but there are some flaws from earlier studies that remain uncorrected: The randomization check is the most serious of these. The fact that a co-author serves as the sole subject in the second experiment makes one uncomfortable as well.

Again, quite apart from the parapsychological nature of the paper, it is my judgment that it would not be accepted for publication in a good psychological journal in its present form.

Schmidt, H. (1973): "PK tests with a high-speed random-number generator"

In this research, Schmidt examined PK using a two-speed fast REG driven by electronic noise (since radioactive material manifesting high-speed decay was not available). Schmidt suggests that a high-speed REG might increase efficiency and allow subjects to identify states of optimal psi readiness and then learn to cultivate these states.

An exploratory study, involving four subjects, including Schmidt himself, was made. Because of the exploratory nature of the study, those results will not be discussed.

In the "confirmatory" study, there were ten subjects, including Schmidt, chosen from a pool of subjects. Subjects were given feedback following each run; and subjects chose in advance to take their feedback in either a visual or an auditory mode. In the former, feedback was in the form of the deflection of a 10-pen recorder, and the subject's task was to attempt to get the pen to go in the target direction. (Deflection from the midline indicated cumulative excess of hits over misses; momentary movement indicated momentary scoring rate). In the auditory feedback mode, feedback was in the form of clicks in a pair of stereo headphones; a click to one ear indicated a hit, while a click to the other indicated a miss.

There were two presentation speeds, either 30 events per second (run = 100 in confirmatory study; duration about 3 minutes) or 300 per second (in which case a run consisted of 1,000 events in the confirmatory study, and this took about the same length of time as for the slower speed). When both speeds were used, the speed alternated from session to session. In the visual condition, the subject could not discriminate speed, and no effort was made to inform the subject unless the subject asked. All three visual feedback subjects participated in both fast and slow conditions. In the auditory condition, one could discriminate between fast and slow, and so subjects were allowed to choose their speed; two chose the fast speed, four chose the slow speed, and one (Schmidt himself) chose to work with both speeds.

Randomization checks: The generator was left unattended for long periods, usually overnight; the numbers of binary 1s and 0s were counted, as were the number of flips. Schmidt indicated that, "depending on whether the experimenter had set the +1 or the −1 as the goal, this number was shown in the display as a 'hit' and the other number as a 'miss' " (p. 108).

Schmidt said that each goal was used equally often. No indication is given of whether or not the target was alternated from session to session or whatever; in any case this would not change the bias problem.

The report states that *the subject's momentary efficiency was frequently rechecked in warm-up runs before they were allowed to contribute to a test session.*

Security: Unspecified—it is not clear if the subject was alone with the apparatus during the testing.

Recording: In the confirmatory series, recording was done manually, but in addition, it was also presented on the pen recorder (visual feedback), or tape recorder (auditory feedback), which allowed for later rechecking. Both procedures, however, allow recording errors.

Results: In the confirmatory series, it was decided in advance to complete 200 runs under each of the four conditions (auditory slow, auditory fast, visual slow, visual fast). Individual subjects, as is typical in Schmidt's work, contributed various numbers of trials. The combined results (hit rates) were as follows:

	Slow	Fast
visual	51.9	50.36
auditory	51.4	50.39

All four hit rates are significantly different from zero. (Although multiple z-tests are performed, the z's are so high that one need not worry about the effect of multiple testing on the Type I error rate.) There was no significant difference between visual and auditory modes, but the hit rate was significantly higher with the slower rate of target generation. This is consistent with the notion that periods of short-term bias were selected by the immediate testing prior to each test trial.

Evaluation: Schmidt found that subjects performed more poorly with the fast rate of target generation. This is consistent with the notion that successful subjects might owe their success to exploitation of short-term generator biases; since subjects were given repeated pretests before each test session to ensure that they were in a good psi state, then, presumably, if one does not beg the question and assume that psi is in operation, the indication that the subject is in a good psi state is equivalent to an indication that a period of short-term bias has been encountered. Suppose that the average length of a "biased" portion (perhaps a warm-up effect) of the target series is N targets, regardless of speed of generation. In the slow speed, the subject is exposed to 100 targets per run and, in the fast speed, to 1,000 per run. Therefore, the subject, if there is a bias, is likely to run over that bias in the 1,000-target run and get a percentage of hits that is lower. In the fast speed of 300 per second, it also means that the time taken between pretest and test may "use up" more of whatever momentary bias exists.

Overall Judgment

Again, because of the fact that the entire claim for anomalousness lies in the departure from randomness in the target series, one cannot accept the results of this experiment as indicative of an anomaly because one cannot be certain that the results are anything more than the consequences of a random generating system with short-term biases. In addition, one might add that once again Schmidt is careless in his arbitrary assignments of subjects and in using himself as a subject, security measures are not discussed in this paper, recording errors are possible, and, finally, the report is rather poorly written. It would not be accepted for publication in any good psychological journal, quite apart from its parapsychological nature.

Schmidt, H. (1974): "Comparison of PK action on two different random number generators"

Here, Schmidt's idea was to use two different REGs, one simple (binary, driven by radioactive decay) and the other complex (strings of 100 binary digits are generated rapidly and if a string contains an excess of +1s, a +1 target is generated, while if there is an excess of −1s, a −1 target is generated. No target is generated in the case of a tie.)

The subject sat facing two lamps, one marked "heads," the other "tails." One or the other was the target on a given trial (although we are not informed how this decision was made), and so the subject's goal was to try to have that lamp light up. Once the subject activated the trial, a random process connected one or the other of the two generators and then, depending on the generator output, one or the other of the lamps was lit up. The subject was provided with continual feedback via a pair of counters indicating number of trials and number of hits.

As usual, Schmidt was the sole experimenter. There was a pilot study in which four subjects, including Schmidt, participated. In the "confirmatory" series, there were 35 subjects, again including Schmidt himself. The subjects were divided into three groups, or "sections": groups 1 (five subjects) and 2 (ten subjects) were composed of members of the parapsychology institute where Schmidt worked, as well as of others who had previously shown good scoring rates. Subjects in the first group knew the goal of the test, while those in the second group, with one exception, did not. Group 3 comprised twenty visitors to the institute who, in a short preliminary test, produced good scores. It was decided in advance to run 1,000 trials for each group. (Actually about 10 percent more were run for each group, but this apparently did not affect the outcome.) Therefore, each subject in group 1 was to do about 200, in group 2 about 100, and in group 3 about 50. (One must ask why these figures were approximate rather than prespecified and exact.) Subjects were allowed to make their responses over one or a number of sessions. Sessions were frequently interrupted by coffee breaks, walks on the porch, or conversation, at the subject's whim. Note that there is no indication of why the number of trials per subject vary so much, or how it was determined when to stop a given subject. Presumably, given that the subject could interrupt or stop the session at any time, it was stopped by the subject (or experimenter) whenever he or she felt like it. Again this is a problem if there are short-term biases in the generator output.

Randomization checks: None were mentioned in the report. The recording of output from the simple generator when it was inactive presumably served as a control, but Schmidt does not suggest this. The determination of whether the simple or complex generator would be active was determined by a random series recorded in a prepunched tape.

Recording: A pen recorder was used to record (*a*) which generator, S (simple) or C (complex), was active; (*b*) the output ("head" or "tail") of the active generator; (*c*) the output of the S generator when it was the inactive generator, and (*d*) the target side (i.e., left lamp or right lamp). Note that the number of hits was not automatically recorded but had to be calculated by hand from the pen recording or read from the counters and then processed.

Security: The experimenter was in the room with the subject, but did not look at the lamps; the recording and generating equipment was one floor down. (Was it overseen by someone else? The report does not say.)

Results: In the confirmatory series, statistically significant PK effects were found for both generators, and there was no significant difference between them. There was no indication that some subjects were more successful than others. No PK effect was found for the S-generator when it was not in active use. Overall, the hit rate was 55.3 percent for the S-generator and 55.3 percent for the C generator. The S-generator when not in active use produced a hit rate of only 50.7, not significantly different from the 50 percent chance rate.

Evaluation: Several errors in text and tables make for difficult reading. That aside, there are a number of concerns:

1. It is unfortunate that Schmidt viewed the recording of S-inactive only as a way of checking for displacement effects; had he recognized its potential as a control, then no doubt he would also have recorded the output of the C-generator when it was inactive, and he would thus have had a control procedure very close to what Hansel has called for— a pair of random events for each trial, and a random selection of which is the target and which is the control. Why was not the output of C recorded when it was inactive? We know from earlier studies that Schmidt has a ten-pen recorder; it seems bizarre that he should not record C-inactive. Did he choose not to report it?

2. In approximately half of the trials in each session, the "head" lamp was the target, and the "tail" lamp was the target for the rest. No indication is given of how this choice was made. If the choice was random, then there should be no problem, but if there was a long string

of one lamp as "head" followed by a long string of the other, this allows the possibility of short-term generator bias to be a problem. However, the lack of above-chance scores for S-inactive apparently nullifies this concern.

3. Again, as seems typical with Schmidt, there was considerable sloppiness in allocation of subjects and trials. As in some of his other experiments, subjects were chosen from a larger pool on the basis of success in preliminary tests. (How are we to be certain that no data were eliminated after the fact? This is a temptation to every graduate student: "Gee, Smith's results throw the whole thing off—but then he did say he had heard somebody mention having been in the experiment, so he wasn't really naive and we cannot really count that!") As usual, Schmidt combines varying numbers of trials from various subjects. This time, having preset the number of trials per group at 1,000, and stating that each subject would participate in about 200 trials, it turns out that the range in the number of trials per subject in group 1 was from 195 to 263, while the range in group 2 was 99 to 125 (when about 100 each were planned) and in group 3, 27 to 124 (where about 50 each were planned). Again there is the worry that this makes exploitation of short-term biases possible; after all, the subject has continual feedback, and can start and stop apparently at whim.

Overall Judgment

Because of the presence of the scores for S-inactive, these data carry more weight, in a sense, than previous data. However, one would wish, before concluding that a genuine anomaly is occurring, to redo the study recording both generators when they are inactive, and taking steps to ensure that there is no post-hoc elimination of subjects.

One would like to see a Schmidt-type experiment with equal numbers of subjects in the various groups, with each contributing the same number of trials, with no feedback, and with adequate controls (using the Hansel-Hyman procedure) to protect against short-term bias. Since Schmidt would no doubt argue that all but the last of these would vitiate conditions necessary for the subject to remain relaxed and well motivated, I would settle even for the last of these measures, for it is by far the most important.

Once again, because of the lack of adequate control procedures, one cannot take the results of this experiment as evidence for the existence of an anomaly. Once again, then, this paper would be unlikely to be accepted for publication in a good psychology journal, quite apart from its parapsychological nature.

Schmidt, H. (1976): "PK effect on Pre-recorded targets"

With this report, Schmidt shifts his research direction toward the study of *retroactive* PK effects. The basic idea was to see whether or not a subject's PK effects could influence a series of randomly generated events that were generated in the past. Three experiments are reported in this paper.

First Experiment

The first part of the first experiment is a conventional PK experiment, using an electronic random generator with $p = 1/64$, $q = 63/64$. For each test, the generator was automatically activated at the rate of 10 trials per second until a hit was obtained, and then the generator stopped, and a hit was indicated to the subject as a weak click in his or her earphone. On average, it should take 64 trials before a hit is obtained, if only chance is operating.

The second part of the first experiment was the same as the first, except that the PK targets consisted of a random sequence of hits and misses generated and recorded on audiotape earlier in the absence of subjects and experimenter. No one saw or heard the sequence until it was played during the experiment. One-half of the sequence was used as the target, while the rest was used as a control and analyzed directly by the computer: A first sequence of 60 runs was generated and labeled A; a second sequence of 60 runs was generated and labeled B. Ten such pairs of sequences were generated. A randomizing procedure was followed to determine whether or not A or B was used. After the first A, B pair was generated, the first subject was selected and tested. After three such subjects were tested, this procedure was repeated: a new A, B pair of 60 runs was generated, 3 new subjects were selected, etc. Results were computer evaluated from paper tape. Subjects were given cumulative feedback immediately after a session.

Confirmatory study: Subjects comprised a prespecified number of laboratory visitors (30). Each subject participated in one session of 20 test runs each, run as two sets of 10 runs each with a 10-minute break in between.

Recording: The equipment was set to run automatically once started, and data recording was automatic.

Randomness checks: Schmidt reports that randomness tests between sessions were consistent with the theoretically expected average run length of 64. It would appear that the control and test runs were of the same length. The conclusion about the lack of bias was based on 4,000 runs.

Results: In the first part of the study (real-time), the average wait for a hit was 55.34, significantly lower than what was expected by chance. In the second part of the experiment, using prerecorded target sequences, there was also a significantly reduced run length of 56.37. Indeed, the results were pretty much the same regardless of whether the targets were generated in the past or the present, which is not inconsistent with the possibility that all the results were due to artifact.

Second Experiment

Here Schmidt used a fast binary random generator that produced sequences of heads/tails at either 30 or 300 per second (as in Schmidt 1973). The subject was given both auditory feedback (variable frequency sound, the goal being to lower frequency) and visual feedback in the form of a fluctuating display needle (the goal being to move the needle to the left). Twenty subjects chosen from a pool of 30 volunteers each contributed two test runs, carried out in a single session. Each run consisted of 4 sequences, and each sequence comprised 256 binary digits.

Two types of targets were used: the target alternated between a "momentarily generated" (or "real-time") target and one from a prerecorded series. There were 4 target sequences of 256 binary bits each; the 128 odd-numbered targets were taken from a predetermined random sequence stored in the computer memory, while the 128 even-numbered targets were provided by triggering the binary random generator. However, while the momentarily generated targets were freshly generated each time, the same prerecorded series was repeated in the same order in each of the four sequences. Following each sequence, there was an intermission of 15 seconds during which feedback was given.

Recording: Recording was automatic. The number of heads and tails was recorded on paper tape before and after runs to ensure that the contents of the memory had not been altered psychically. (That *would* be strange; but Schmidt must think that it is easier to go back in time and affect quantum processes than it is to alter the data stored in a computer.)

Randomness checks: There were five control runs before and after each test session; on each run, 128 numbers were generated, stored in memory, retrieved and recorded. Over the 200 control runs, there were 12,715 heads and 12,885 tails; the CR (critical ratio) = 1.1, which is not significant. As an additional control measure, the target side of the generator was alternated on each session in order to protect against a constant bias.

Results: Compared to a chance rate of 50 percent, the hit rate for momentarily generated targets was 50.815 ($p < .05$ one-tailed), while for prerecorded targets, it was 52.95 percent ($p < .0005$).

Schmidt concluded that repeated feedback of prerecorded data leads to increases in scoring rates: the within-subjects difference between prerecorded and momentarily generated targets was significant. He assumes that the superior rate obtained for the prerecorded targets was due to repeated presentations.

Evaluation: Again, there is reason for concern about the adequacy of the randomness checks; the Hansel procedure would control for short-term biases; the procedure followed here does not.

Third Experiment

In this experiment, there were two binary random generators, one, the "difficult" one, with $P(\text{hit}) = 1/8$ and the other, the "easy" one, with $P(\text{hit}) = 7/8$. These two were activated in a random sequence so that the subject never knew which would produce the next target. In the pilot study and in the first half of the confirmatory series of this experiment, feedback was presented in the form of clicks to the left ear (hits) or right ear (misses) via stereo headphones. Then, and no reason for this is given, the rest of the confirmatory series employed auditory feedback that took the form of either a high or a low tone.

The random order of presentation of generators was determined by a prerecorded random sequence of 64 binary digits ("heads" and "tails") produced by a random generator (having $P(\text{head}) = P(\text{tail}) = 1/2$) prior to the experiment: a head led to the use of the 7/8 generator, while a tail led to the use of the 1/8 generator. Since more heads lead to increases in the total score, there should be motivation to want the 7/8 generator, and this might lead the subject to influence the prerecorded sequence by means of PK. Each sequence of 64 binary digits was used 4 times in succession, so that each head or tail caused four triggerings of either the easy or the difficult generator. There were 40 sessions in the confirmatory series, and each of these consisted of four test runs; each run involved 64 random events, and feedback was given after each run. Twenty-eight volunteers participated, and some of them performed in up to three sessions because of a shortage of volunteers.

Results: While the pilot series seemed to support the existence of a PK effect, Schmidt said, the effect in the confirmatory series was only "marginally significant" at the .05 level. (Note: The abstract says that the PK effect on prerecorded sequences was confirmed at the .05 level. Yet the t-value is only 1.3; this is only the .10 level *one-tailed!* However, Schmidt also provides a CR [z-score] of 2.03, and this is presumably what he refers to as "marginally significant" at the .05 level. He only gives the t-level in the table; why has he been using ts instead of zs, and why does he only refer, it seems, to the z in the abstract? Why does he use both z and t in the table for the first experiment?)

Evaluation: The third experiment was a failure; apart from a significant deviation in a pilot series, there is nothing, Schmidt's born-again CR notwithstanding.

Overall Judgment

These are the most automated and least sloppy of Schmidt's studies so far. The number of subjects and number of trials were prespecified; automatic equipment operation and recording was used. Control runs appear to have been of the same length as test runs, and were interspersed with them, although it seems that the decision that there was no bias was based on the overall set of control runs for a given experiment, and no attempt was made to check for trends that might cancel each other out but still give short-term biases that could aid the subject.

Oddly enough, just as this writer begins to be more content with the procedures (although not yet persuaded that biases in the random sequences have been ruled out; the Hansel procedure would be desirable for that purpose), Schmidt begins to sound more conservative:

> With any interpretation of the results one has, of course, to be cautious for several reasons. First, one might want to postpone serious discussion until we have more detailed experimental information from several independent experimenters. Second, the failure of the third experiment to give more than a marginally significant PK effect reminds us that we may still be overlooking some vital factors which have a stronger effect on the test results than the variables we are studying. . . . (pp. 290-291)

I concur.

Schmidt, H. (1978): "A take-home test in PK with prerecorded targets"

First Experiment

The basic idea here was to generate, using a binary random generator, a sequence of tones on an audiotape, the pitch growing successively higher for each hit and lower for each miss. Each run comprised 512 binary events, and there were 12 runs per tape. A random assignment procedure was used to categorize each tape as either high or low. Then the subject would take home a *copy* of the tape, and as he or she listened to it, the attempt was made to increase the tone if the tape was assigned to the high category, or to decrease the tone if the tape belonged to the low category. Subsequently, Schmidt, again the sole experimenter, would evaluate *his* copy of the tape for imbalance between the numbers of hits and misses. Actually, he examined a computer printout of hits/misses prepared at the time the tape was prepared, but which had not been viewed until after the subject had finished his task. (Thus, this printout was, effectively, the data.) There were 12 runs each of high and low, and each run consisted of 512 binary digits.

Second Experiment

The second experiment was similar to the first, except that, whereas in the first experiment, the 10 subjects were able to play around with the machine before taking home the prerecorded tape, in the second, 64 volunteers were contacted by telephone and the tapes mailed out. In addition, half the runs were "group" runs, in that four different subjects received identical recordings; the other half were individual runs. (One side of each subject's tape was an individualized series, while the series on the other side was sent to three other subjects as well.) Further, for half the subjects in the individual and group tests, the printout data were seen by an assistant, who did not know their meaning, before the subjects made their PK efforts. There were 12 individual runs and 12 group runs; half of the runs in each condition were low and the rest high. Thus there were 64 individual runs and 16 group runs for a total of 80 runs.

Randomization checks: None at all were reported for the generation of the tapes.

Recording: No data needed to be recorded. All that was required was to analyze the original computer printout for departures from randomness. This was presumably done manually.

Results: For experiment 1, all we are told is that the CR = 3.34 ($p < .001$); no data are presented. We are also told that a t-test on the 10 total scores (comparing each subject's scores for high and low tapes) yielded a $t = 0.86$, 9 df. This difference in score was nonsignificant.

There were no significant findings in experiment 2. However, Schmidt reanalyzed the data and found that the squares of the CRs (in effect, z-scores squared) were all above the chance level of 1, and this is taken to suggest a combination of PK hitters and missers. Combining all CR-squared (80 contributions) gives an average CR-squared of 1.31, $p = .03$, only "suggestively high."

Evaluation: No data were provided in this paper, just test statistics.

Although a significant effect was found in the first experiment, the second really serves as a failure to replicate, although, because of the changes in procedure, Schmidt takes the outcome to suggest that the informality of sending out the tapes by mail in some way decreased the likelihood of psi.

The absence of randomization checks (none were reported) by itself would render this paper unacceptable for publication in psychology journals.

Terry, J., and H. Schmidt (1978): "Conscious and unconscious PK tests with prerecorded targets"

This study is a follow-up to the previous one, and this time subconscious as well as conscious PK efforts were under scrutiny. There were two experimenters involved.

There were three separate experiments of 20, 20, and 30 sessions, respectively. In the first two of these, there was a different subject for each session, while in the third, all subjects participated in more than one session, and indeed some worked simultaneously on the same PK task.

For the "conscious" runs, high tones were presented at random time intervals and the subject's task was to increase the number of such sounds. For the "subconscious" task, high- and low-pitched sounds were presented at random time intervals, and the subject's task was to react to high-pitched sounds by pressing a switch as quickly as possible, but not reacting to low-pitched sounds. It was expected that the intense concentration on the sounds might lead to subconscious PK efforts to shorten the interval between successive sounds.

One experimenter used a randomization routine on a pocket calculator to decide which of two tapes would be used as an experimental tape and which as the control. Without knowledge of this decision, the other experimenter, Schmidt, prepared a pair of digital cassette tapes, each containing six sequences, corresponding to six runs, of random numbers. The numbers were 0, 1, and 2, chosen so that the relative frequency of 0 would be 15/16, while the relative frequencies of 1 and 2 would be 1/32 each. For the "conscious" runs, the 1s produced a high tone, while for the subconscious runs, the 1s and 2s produced high and low tones, respectively.

The control tape was not examined until the end of the experiment, at which time it, along with the experimental tape, was read and analyzed automatically by a computer.

The authors presented the combined results of the three experiments, and the only significant results were as follows: There was significant PK missing ($p < .005$, 2-tailed) in the conscious runs, as well as a significantly high variance for those runs ($p < .005$, 1-tailed). There was not a significant difference between the 70 conscious PK sessions and the 70 corresponding control sessions. However, it was found that the control sessions were biased toward a smaller number of high tones, although the bias was nonsignificant. The authors added:

> This non-significant bias, which also appears in the subconscious runs and the corresponding control runs, raises the question whether perhaps the random generator was biased. An extensive randomness test at the completion of the experiments, however, indicated no such bias. (p. 40)

Yet again the randomness test was one hundred times longer than the length of the series generated and recorded on the tapes.

Evaluation: In this experiment, even the experimenters suspect nonrandomness of the generator. Given that they were prepared to accept psi-missing as well as psi-hitting as evidence of PK, all that was necessary to produce a significant effect was to initially generate two tapes that differed in some way because of generator bias. That is precisely what appears to have happened.

This study would not be acceptable for publication in a good psychological journal, quite apart from its subject matter.

Schmidt (1979a): "Search for psi fluctuations in a PK test with cockroaches"

This is a brief report in which Schmidt suggests that the reason that cockroaches were unable to avoid shock in his 1970c study may be that they do not encounter such shocks in nature and therefore have no preparedness to avoid them. By using random delivery of shocks at two different probabilities, one might have a more sensitive measure of their

PK abilities, he argued, and by using repeatedly the same recorded series of events, again this might aid the cockroach in coming to be able to reduce the number of shocks. He thus compared the actual shock rates when the a priori probability of a shock was 1/4 against when it was 3/4. No evidence for PK was found. Presenting the same series of recorded random events 32 times did nothing to improve the avoidance of shocks.

Schmidt also reported in brief his failure to find PK effects in studies with algae, yeast, and wingless fruit flies.

Schmidt, H. (1979b): "Use of stroboscopoic light as rewarding feedback in a PK test with prerecorded and momentarily generated random events"

In this study, subjects attempted to affect mentally the duration of time intervals determined by random processes. The length of the first part of each interval was determined in advance of the test; the second part was determined at the time, so that prerecorded and momentarily generated random events were both involved, and the main goal of the study was to compare the two with regard to PK effects.

During *on* intervals, the subject was exposed to a strobe light flashing at a frequency that he or she had preselected as being particularly pleasing, while during *off* intervals, the subject viewed a practically constant light source. The subject was instructed to try to lengthen the *on* intervals and shorten the *off* intervals.

A test run consisted of 8 *on* and 8 *off* intervals. Each time interval, *on* or *off*, had two sections; each section was n units (a unit = 5/16 seconds) long. Before a run, 16 random numbers were generated by means of an electronic die (based on radioactive decay): the random number n was determined by the number of "rolls" of the electronic die before an 8 came up. The second part of each interval was determined *during* the run: the electronic die was activated once after every time interval during this run, and the section was terminated when an 8 appeared. The subject could not sensorially distinguish between momentarily generated and prerecorded sections.

There were 200 test runs divided among the 12 subjects. The report does not indicate whether subjects were given feedback. *Subjects first made one or two unrecorded trial runs; if they still felt good about the test, they* were then allowed to contribute to the test runs. When a subject returned for another session, he would always begin with one or two warm-up runs after which it was decided whether test runs should be undertaken or not.

Randomization controls: None mentioned.

Security: Not stated.

Results: No data were provided, not even means; we are only told that the observed *on* intervals were 7 percent longer than would be expected by chance, and that this is statistically significant (CR = 4.26) at the $p < .0001$ level. The *off* intervals were a nonsignificant 0.5 percent shorter than chance expectation. There was an equal effect on prerecorded and momentarily generated intervals (6.8 percent, CR = 2.9, and 7.3 percent, CR = 3.1, respectively). So both prerecorded and momentarily generated *on* intervals showed the effect, but there was no effect for *off* intervals.

Evaluation: No data or discussion is offered regarding randomization tests; based on past history, one must suspect that there were no checks for short-term bias. The warm-up procedure, whereby the decision to proceed to test trials depended on the results of the warm-up, furnishes an excellent opportunity to select for short-term bias if it exists.

Given that data are not reported, even in summary form (means, variances), one obviously cannot evaluate the data. We do not even know how the 200 test runs were divided among the 12 subjects, except that it had to be unequally. Did one subject have a number of runs in a stretch and contribute disproportionately to the data? We cannot know. Note that if it is a question of bias, the *on* and *off* intervals were interspersed; why should the bias only affect the *on* trials? That is why we should like to see the data for each subject, to see whether there is a pattern: Did all subjects, for example, show an increase in *on* length but not in *off?*

Overall Judgment

This study is too sketchily presented for the evidence to be evaluated in any meaningful way. It was a conference paper, which accounts no doubt for the lack of detail, but as far as I know it has not been published in any other form. However, it seems that the same procedure that allows for selection of generator bias, and which has been criticized in the discussions of some of the earlier studies, was in operation here.

It would not be publishable in a good psychology journal, quite apart from its theme.

Schmidt, H. (1981): "PK tests with prerecorded and preinspected seed numbers"

In this research project, Schmidt addresses the question of what would happen if, instead of using individually generated random numbers, one uses a causal algorithm that begins with a randomly generated seed number. Will there still be a seed effect? If so, does it matter if a human observer sees the seed number in advance of the test?

Each run consisted of 512 target numbers that were generated by an algorithm that used a seed number derived from a truly random process. Seed numbers were obtained in advance, and half of them were inspected by the experimenter (". . . theoretically, the experimenter would have been able to calculate from the seed number the PK target sequence and the final run score" [p. 88]).

The task involved a circular series of 16 lamps in which only one lamp is lit at any given time. Random-number generation worked on a modulus-16 basis, and with each generation of a random number, the light moved in one step either clockwise or counterclockwise: it would move clockwise until a 3 occurred, at which time it would begin to move in the opposite direction for each number generated until a 12 was obtained. This process continued until 16 clockwise-counterclockwise pairs were completed.

The subject's task was to try to make the light move clockwise. (There was an inverter switch that the subject could use if he or she wanted to make the light move counterclockwise; the switch simply interchanged clockwise and counterclockwise movement in the display.) Auditory feedback was also provided via a decaying gong sound which was on as long as the light was moving in a clockwise direction. The random generator operated as a 16-sided die, and left on its own the light should move an average of 16 steps in the same direction before changing. Essentially, then, the generator produced a series of 32 random time intervals and the subjects's task was to lengthen the odd-numbered intervals and shorten the others. At the end of 16 pairs, display counters gave the number of hits (clockwise) and misses, respectively. As Schmidt pointed out:

> Thus, by choosing a favorable seed number we can pick out a section of the number chain which is favorable for success in the experiment. And since each section used in a test run comprises only about 1/1024 of the total chain length, there is sufficient opportunity for such favorable sections to occur. (p. 91)

(It is interesting to note that that comment closely relates to what I have been calling attention to with regard to the problem of free-play and optional starting!)

The subjects, as is so often the case in Schmidt's experiments, were a kind of ragtag lot. There were two groups; there were 11 subjects in Group U, which was predominantly *unselected* (why only predominantly, we are not told), and 4 subjects in Group S, S indicating that these were subjects who had been *selected* because they appeared to be particularly promising. One of them "was a newcomer who impressed the experimenter with his confidence and his proficiency in martial arts" (p. 94)!

In the main experiment, there were 100 test runs with Group U, and another 50 runs with Group S. Subjects could participate in more than one session. Prior to running the subjects, two blocks of 100 and 50 random numbers (S and U, respectively) were generated using radioactive decay, stored in a computer, and printed out on the computer. A template was attached to the printer so that Schmidt could only see the seed numbers in the odd-numbered columns; these he read aloud; he then closed his eyes and tore off the paper and put it in an envelope for storage. (What about clairvoyance?)

In play sessions, random numbers were obtained by activating the generator in the machine; for the other sessions, the algorithm was fueled by a seed number from the memory block; no seed number was used twice.

Most sessions began with play runs. After the play runs, a *flexible number of sessions* (on average, 8) were conducted. *The decision when to end a session was "made rather subjectively, depending on the experimenter's mood and confidence. In a few cases where the subject appeared uncomfortable and prone to psi-missing under all conditions, the session was terminated. after the play runs."* This of course is a subtle form of data selection. Subjects who seem incapable of doing well are eliminated.

Security: Not mentioned.

Randomization checks: None mentioned.

Results: Schmidt analyzed all runs together, for each group, and then separately analyzed only the preinspected runs. He reported significant departures from chance expectation for both the S group and the U group ($p < .0005$, $p < .05$, respectively). The PK effect was found with both the preinspected and the uninspected seed numbers for the S group; although the departures for all runs and for only the preinspected runs were in the same direction for the U group, only that for all runs was significant.

Evaluation: The quality is in many ways improved over earlier studies, but there are still the basic problems of randomization checks (none mentioned) and optional starting following free play. Multiple zs increase the likelihood of Type I error above the stated levels, and given that the zs (or CRs, as Schmidt calls them) are not very big, this is of serious concern as well.

Overall Judgment

Once again, this paper would be unlikely to be accepted for publication in a good psychology journal, quite apart from its parapsychological nature.

Schmidt, H. (1985): "Addition effect for PK on prerecorded targets"

While this study receives points for imagination and creativity, it is very poor from a methodological viewpoint. Schmidt's research question in this instance is, What happens when two consecutive PK efforts are made in the attempt to influence the same prerecorded binary events? Do the two efforts contribute equally to the outcome, or does the first have the stronger effect, in line with Schmidt's "quantum collapse model." There are three experiments included in this report. In the first, Schmidt is both experimenter and sole subject. In the other two, PK efforts are made successively by two subjects, a test subject and a control subject. The control subject attempted to influence the binary events in some cases in the same direction as the test subject, and in some cases in the opposite direction. Schmidt, again the sole experimenter, *is also the test subject in each case!* Schmidt writes:

> I had decided in advance to base the final conclusions . . . on the performance of the test subject only, because neither of the control subjects had previously shown interesting scores. (p. 238)

He found that when he went first, the cumulative PK score increased, while when he went second, there was no significant difference from chance. These results, he concluded, support his quantum collapse model.

Randomization checks: None described.

Evaluation: Given that Schmidt is both experimenter and subject, again this should be viewed as nothing other than a pilot study.

Overall Judgment

This is a poorly executed experiment. It would have no place in a good psychology journal, quite apart from its parapsychological theme.

Schmidt, H., R. L. Morris, and L. Rudolph (1986): "Channeling evidence for a psychokinetic effect to independent observers"

This study is described by its authors as "an experiment in psychokinesis, performed under unusually tightly controlled conditions. In particular, the experimenter was supervised by observers from another laboratory such that negligence and even fraud by the experimenter could be ruled out as an explanation of the observed effects."

It involved a PK test using prerecorded random seed numbers in an attempt to channel psi evidence from Schmidt's lab in San Antonio to two independent observers (Morris and Rudolph) at Syracuse University. There were a number of subjects involved: Schmidt himself contributed a substantial (unstated) number of trials. There were ten sections of test runs; Schmidt was the sole subject in four of them, and one of two subjects in three others. Another involved a single subject, another involved two subjects, and a third involved six.

The methodology was as follows:

Appendix 1

1. Schmidt used a computer with a geiger counter to generate and print a supply of truly random six-digit seed numbers.

2. At the start of each section, Schmidt decided on the number of test runs to be made and which of three test arrangements to use. (The three arrangements were SWING1 and SWING2, two computer programs that displayed a pendulum on the monitor whose amplitude was governed by the series of 0s and 1s provided by the sequence generated by an algorithm using the seed number, and CLICK, a program that provided auditory feedback instead of visual. SWING1 was used twice, SWING2 was used six times, and CLICK was used twice.)

3. Schmidt then sent a corresponding number of seed numbers to the first observer, O_1, and also informed O_2 by telephone of the number of trials and the test arrangement. O_2 had already received either the corresponding computer (!) or the program to calculate the scores for the seed numbers.

4. O_2 used his own electronic noise REG to generate a sequence of binary decisions (H or L); and for each pair of seed numbers, the first was labeled on the basis of this number, and the other was the opposite.

5. Schmidt obtained the list of H-Ls and then gave subjects a list of seed numbers along with the target assignments and a computer so that they could work at home at their convenience. The subjects were instructed to try to influence the outcome so as, in effect, to produce a greater correlation between the H-L assignments and the order in the pairs of seed numbers.

6. Once a section was completed, the observers were informed that they could now use their computer to calculate the scores for the seed numbers.

There were various (prespecified) numbers of trials in each section, but in each trial, a pair of seed numbers was the basic unit, and one calculated the difference between the score for the H number and the score for the L number to produce the basic datum.

Randomization checks: None were really needed, in that comparison is between H and L within each pair. Only if in some way there were a correlation between the two randomization procedures (that yielding the seed number and that yielding the H-Ls) would there be a problem, and it is difficult to imagine such a correlation.

Detailed Critique of the Schmidt Studes   169

Results: If there is no departure from chance expectation, then we should expect that the mean pair-score does not differ significantly from zero. The authors used a rank-order test and calculated a z-statistic for each of the ten sections. Eight of the ten did not differ significantly from expectation, and the other two did, but only at the .05 level, one-tailed. When all the z-scores were combined, it was found that the resultant z was significant at the .003 level. I was able to obtain the raw data and when the entire data set is analyzed, with no regard for its division into ten subgroups, a significant overall deviation from what would be expected by chance is still evident; $z = 2.2$, $p < .02$. (Note: The ten sections could be taken as ten independent attempts to demonstrate the effect, only two of which were significant and then only at the .05 level).

Evaluation: Although the effect is not all that large, this study is so much better designed and executed than the earlier studies that at least one has to concede that replication attempts should be made. This is not to say that psi has been demonstrated, for even the authors speak only of an "anomalous correlation," and in that spirit, one should postpone any further interpretation or judgment *until* replication by independent experimenters can be produced.

Nonetheless, although the authors show more circumspection than Schmidt has done in the past when acting alone, and although they indicate that really all they mean by psychokinesis in this case is an "anomalous correlation, not understandable in terms of current physics" (p. 18), clearly they are suggesting that either the subjects or the observer who calculated the random sequence of H and L pairs had some kind of mental influence on the outcome; otherwise, why use subjects at all? If one begins, instead of begging the question by assuming that such mental influence exists, by assuming that the "effect" has already been produced once the H and L pairs are assigned, one has a better handle on the mystery. Otherwise, if one is to consider that the subject may have had an effect, one runs into circularity immediately. Since it is agreed that the algorithm is inviolate, so that a given seed number always generates the same score, and since it is also assumed that the H or L assignment does not reverse itself as a result of the subject's actions, then what can change? Only the seed number itself, which was selected by a truly random process.

The implication is that the subject, through what appears essentially to be "wishing," is able to influence a subatomic process that occurred at some time in the past in order that the emission of a particle will generate a six-digit number that will bear some desired relation to another six-digit number. However, the subject cannot know the nature of this desired relationship, for these numbers are merely seed numbers that ultimately generate *other* numbers, which are themselves the subject of interest; the order of a pair of seed numbers is not directly related to the order of the pair of resulting numbers, and it is this latter order that is of interest. So the subject's task is an extremely complex one, one that the human mind by itself would be unable to master, but one, we are asked to believe, that is somehow accomplished, at least on some trials, by the marvelous facility of simply wishing for it and leaving the rest to psi.

Although this particular piece of research is much better in design and execution than any of the previous studies, nonetheless it is still seriously marred by methodological carelessness and unnecessary complexity. Why, for example, use three different tasks, and run varying numbers of trials with each, and then mix all the data together? Why was the use of these tasks not decided in advance, rather than apparently being left to the whim of the experimenter, who informed the observers during the experiment as to what task was being employed at a given time? Why did the experimenter himself find it necessary to be a subject for a substantial proportion of the trials? Why were some sections run with Schmidt alone as subject, while others involved up to six different subjects? Why were ten "sections" run and then blended together? In this particular case, given that the subjects were not in a position to really do anything, except, supposedly, by psychic means, this carelessness in methodology is probably irrelevant, as is the fact that the experimenter, Schmidt, himself contributed a substantial proportion of the data. However, such sloppiness does not generate great confidence in the experimenter.

The complexity is needless; the basic pair-score is the difference between two scores that are themselves the results of an algorithm that is started off by a randomly chosen seed number. Yet the algorithm apparently generates a series of binary 0s and 1s that are used to control the presentation of either the pendulum or the clicks. Apparently 128 such binary bits are used for each trial, and so one might expect that the "score" for a trial would be upper-bounded by 128. However, such is not the case, as use of the BASIC algorithm or examination of the example of two scores demonstrates. It would have been so much easier simply to use one task, to have each subject perform for a fixed number of trials, to use the sum of binary bits as the score, and so forth.

Obviously, it would have made sense in any normal psychological experiment to have a control condition where no subjects were used at all: just analyze the data at this point and compare them with what is generated when subjects are involved. However, in parapsychological research, it might be argued that, even if the controls showed similar significant differences, the differences were brought about by the psychic energies of, say, the observer who used his random generator to generate the target assignments. Indeed, in this report by the authors, this is considered:

> One might even wonder whether the PK effect did not enter through the second observer, subconsciously forcing his random generator into producing favorable target assignments. (p. 18)

That, of course, would involve clairvoyantly viewing the list of seed numbers and, perhaps by precognition or clairvoyance, determining what the resulting score will be once the seed is fed into the algorithm. One would have to do this for each pair of numbers and then, after deciding which resultant score is the higher, bring psychic influence to bear on the electronic noise in such a way that it would cause the necessary binary digit to be generated. This is not only quite a mean feat, but it is done repetitively during the session, although not always accurately, for the deviation from chance although significant statistically is not dramatic by any means.

This process sounds well beyond the ability of the conscious mind, and it would seem unlikely that the "unconscious" mind could do any better. However, modern parapsychological theorizing suggests that psi operates in a "goal-oriented" fashion, that one does not go through the intricate kind of calculation and analysis that I have just adumbrated. One simply desires an outcome, and that is enough.

Overall Judgment

This study is much better designed and executed than earlier Schmidt studies. The observed effect was relatively small, and in the light of unnecessary methodological sloppiness and complexity, one would expect that the authors would want to attempt to run a more refined experiment, with a control group in which no subjects serve, in order to try to understand the cause of the correlation, if it is replicated. It is far too early, in my view, to begin speculating about psi, even though it is not easy to come up with an explanation for the (somewhat marginal) results on the basis of the report.

Appendix 2

Summary of Randomization Checks in Schmidt's Works

Study	Randomization Checks
1969a	Five million numbers were generated by the REG for control purposes, and the frequency of all four numbers and all 16 sequential pairs were calculated and did not differ significantly from chance expectation. These five million were recorded on 100 different days, "preferably directly after the experimental sessions." It may have been preferable to always do such a check immediately before the session, instead of "preferably after." [Note that even that would not be an adequate control. Hansel's idea about generating two runs each time and randomly selecting one for the actual target and the other for a control would seem much better.] There were only 39 sessions, and we are not told over how many days these sessions were spread. We are given no

information about the relationship between the control runs and the tests except for the "preferably after" comment.

1969b The randomness check employed is not described, but reference in this regard is made to an unpublished paper.

1970a Four million numbers were generated on many different days (not specified what was relationship of these days to experimental sessions). Number of +1s, –1s, and number of flips examined. Also divided the numbers into 10,000 sequences of 400 and looked at number of +1, –1, and flips. The outcome of this was consistent with normal distribution. [But what about relationship between 400 sequences? Do a number of high 400s occur in a row, for instance?]

After first half of confirmatory test, the two outputs of the generator were interchanged internally—so any bias would be reversed. (We are not told what the breakdown of scores were before and after this change. Could it be, for example, that the deviation was highly negative pre-change, and positive post-change, to give an overall negative score, but one much less impressive than in earlier studies [1969a, 1969b]?)

1970b No experiments reported here. This is a description of Schmidt's *binary* RNG. He discusses the randomization check:

N^+ = number of generated +1s
N^- = number of generated –1s
F = number of flips

For a random sequence, the variables

$X = (\text{SQRT } 1/N)(N^+ - N^-)$
$Y = (\text{SQRT } 4/N)(F - N/2)$

have independent normal distributions with

Summary of Randomization Checks in Schmidt's Works 175

$$X \text{ mean} = Y \text{ mean} = 0$$
$$X^2 \text{ mean} = Y^2 \text{ mean} = 1$$
$$(XY) \text{ mean} = 0$$

The first randomness check involves counting, for one long generated sequence, N^+, N^-, and F, and checking to see that X^2 and Y^2 are not unduly large. Secondly, one might generate a certain number, S, of number sequences of length N and record the S X-values and Y-values obtained. These should have a normal distribution, which might be evaluated by a goodness-of-fit test.

1970c Cat: 8 hours per night for 24 nights, lamp *outside*, system left running. No bias in 691,200 trials. [Note: Sessions were only ½ hour in length; no short-term checks.] Between sessions, we are told, machine was left running continuously to verify (in some unspecified manner) the lack of bias.

Roaches: RNG left running each night after PK test; no bias found. *Both* generator outputs used equal number of times to produce +1 (shock). (Not explained whether the reversal was made halfway through experiment, or at various times.)

1972 (with Pantas) No mention of randomization checks.

1973 RNG left unattended for long periods, usually overnight; numbers of N^+, N^-, and flips counted. No doublets required, since there are only two states.

P. 108: "Depending on whether the experimenter had set the +1 or the −1 as the goal, this number [was] shown in the display as a 'hit' and the other number as a 'miss.' " He said that each was used equally often. No indication of whether the target was alternated from session to session or whatever; in any case this would not change the bias problem.

1974 No randomization checks mentioned.

1976   Randomness tests at the completion of the sessions were consistent with the theoretically expected average run length of 64.

1978   No randomness checks were mentioned.

1978   (with Terry) Control tapes were found to have nonsignificant bias, yet an extensive randomness test at the completion of the experiments indicated no such bias. However, it was 100 times as long as the random series generated for a set of six runs.

1979   None mentioned.

1981   None mentioned.

1985   None mentioned.

1986   Not required.

References

Akers, C. 1984. Methodological criticisms of parapsychology. In *Advances in Parapsychological Research*, vol. 4, ed. by S. Krippner. Jefferson, N.C. McFarland.
Alcock, J. E. 1981. Parapsychology: Science or Magic? Elmsford, N.Y.: Pergamon.
———. 1984. Parapsychology's last eight years: A lack of progress report. *Skeptical Inquirer*, 8:312-320.
———. 1985. Parapsychology as a "spiritual" science. In *A Skeptic's Handbook of Parapsychology*, ed. by P. Kurtz. Buffalo, N.Y.: Prometheus Books.
Allison, P. D. 1973. Sociological aspects of innovations: The case of parapsychology. Unpublished masters thesis, University of Wisconsin.
Beloff, J. 1973. *Psychological Sciences*. Crosby Lockwood Staples.
———. 1975. Reviews of *Uri Geller: My Story*, by U. Geller, *Uri*, by A. Puharich, and *Superminds: An Inquiry into the Paranormal*, by J. Taylor. *Journal of Parapsychology*, 39:242-250.
———. Historical overview. In *Handbook of Parapsychology*, ed. by B. B. Wolman. New York: Van Nostrand Reinhold.
———. 1980a. Seven evidential experiments. *Zetetic Scholar*, 6:91-94.
———. 1980b. Could there be a physical explanation for psi? *Journal of the Society for Psychical Research*, 50: 263–272.
———. 1982. Psychical research and psychology. In *Psychical Research: A Guide to Its History*, ed. by I. Grattan-Guiness. Wellingborough, U.K.: Aquarian Press.
———. 1984. Research strategies for dealing with unstable phenomena. *Parapsychology Review*, 1:1-7.
———. 1985. What is your counter-explanation? A plea to the skeptics to think again. In *A Skeptic's Handbook of Parapsychology*, ed. by P. Kurtz. Buffalo, N.Y.: Prometheus Books.
Berkowitz, L. 1986. *A Survey of Social Psychology*, 3rd ed. New York: Holt, Rinehart & Winston.

Bierman, D. J., and D. H. Weiner. 1980. A preliminary study of the effects of data destruction on the influence of future observers. *Journal of Parapsychology*, 44:233-234.

Bird, J. 1977. Applications of dowsing: An ancient biopsychophysical art. In *Future Science*, ed. by J. White and S. Krippner. New York: Anchor Books.

Bisaha, J. P., and B. J. Dunne. 1979. Multiple subject and long-distance precognitive remote viewing of geographical locations. In *Mind at Large*, ed. by C. Tart, H. E. Puthoff, and R. Targ. New York: Praeger.

Blackmore, S. 1980. The extent of selective reporting of ESP ganzfeld studies. *European Journal of Parapsychology*, 3:213-219.

———. 1982. *Beyond the Body*. London, U.K.: Heineman.

———. 1983a. Unrepeatability: Parapsychology's only finding. Paper presented at the Parapsychology Foundation Conference, San Antonio, Texas, October 1983.

———. 1983b. Comments on Hövelmann's Seven Recommendations for the future practice of parapsychology. *Zetetic Scholar*, 11:141-143.

———. 1984. A psychological theory of the out-of-body experience. *Journal of Parapsychology*, 48:200-218.

———. 1985. The adventures of a psi-inhibitory experimenter. In *A Skeptic's Handbook of Parapsychology*, ed. by P. Kurtz. Buffalo, N.Y.: Prometheus Books.

Boring, E. G. 1966. Paranormal phenomena: Evidence, specification, and chance. Introduction to C. E. M. Hansel's *ESP: A Scientific Evaluation*. New York: Scribner's.

Bozarth, J. D., and R. R. Roberts. 1972. Signifying significant significance. *American Psychologist*, 27:774-775.

Braude, S. E. 1978. On the meaning of "paranormal." In *Philosophy and Parapsychology*, ed. by J. Ludwig. Buffalo, N.Y.: Prometheus Books.

———. 1979. The observational theories in parapsychology: A critique. *Journal of the American Society for Psychical Research*, 73:349-366.

Bunge, M. 1984. What is pseudoscience? *Skeptical Inquirer*, 9:36-46.

Caulkins, J. 1980. Comments. *Zetetic Scholar*, 6:77-80.

Casrud, K. B. 1984. Out of the frying pan: A reply to Sommer and Sommer. *American Psychologist*, 39:1317-1318.

Cerullo, J. J. 1982. *The Secularization of the Soul*. Philadelphia: Institute for the Study of Human Issues.

Child, I. L. 1984. Implications of parapsychology for psychology. In *Advances in Parapsychological Research*, vol. 4, ed. by S. Krippner. Jefferson, N.C.: McFarland.

———. 1985. Psychology and anomalous observations: The question of ESP in dreams. *American Psychologist*, 40:1219-1230.

Collins, H. M. 1976. Upon the replication of scientific findings: A discussion illuminated by the experiences of researchers into parapsychology. *Proceedings of the Fourth International Conference on Social Studies of Science* (Mimeo). Ithaca, N.Y.: Cornell University.

Collins, H. M., and T. J. Pinch. 1979. The construction of the paranormal: Nothing scientific is happening. In *On the Margins of Science: The Social Construction of Rejected Knowledge*, ed. by R. Wallis. *Sociological Review Monograph* 27:237-270.

———. 1982. *Frames of Meaning*. London, U.K.: Routledge & Kegan Paul.

Cox, W. E. 1976. A preliminary scrutiny of Uri Geller. In *The Geller Papers*, ed. by C. Panati. New York: Houghton Mifflin.

Crandall, J. E., and D. D. Hite. 1983. Psi missing and displacement: Evidence for improperly focused psi? *Journal of the American Society for Psychical Research*, 77:209-228.

Druckman, D., and J. A. Swets. 1988. *Enhancing Human Performance*. Washington, D.C.: National Academy Press.

Dunne, B. J., and J. P. Bisaha. 1978. Multiple channels in precognitive remote viewing. In *Research in Parapsychology 1977*, ed. by W. G. Roll. Metuchen, N.J.: Scarecrow Press.

———. 1979. Precognitive remote viewing in the Chicago area: A replication of the Stanford experiment. *Journal of Parapsycholgoy*, 43:17-30.

Dunne, B. J., R. G. Jahn, and R. D. Nelson. 1983. *Precognitive Remote Perception* (Technical Note PEAR 83003). Princeton Engineering Anomalies Research Laboratory, School of Engineering/Applied Science, Princeton University.

Edge, H. L., and R. L. Morris. 1986. Psi and science. In *Foundations of Parapsychology*, by H. L. Edge, R. L. Morris, J. Palmer, and J. H. Rush. London, U.K.: Routledge & Kegan Paul.

Eisenbud, J. 1976. Review of *The Geller Papers*, ed. by C. Panati. *Journal of Parapsychology*, 40:321-325.

———. 1977. Paranormal photography. In *Handbook of Parapsychology*, ed. by B. B. Wolman. New York: Van Nostrand Reinhold.

Epstein, S. 1980. The stability of behavior: 2. Implications for psychological research. *American Psychologist*, 35:790-806.

Estes, W. K. 1976. The cognitive side of probability learning. *Psychological Bulletin*, 83(1):37-64.

Evans, D. 1973. Parapsychology: What the questionnaire revealed. *New Scientist*, 57:209.

Fishman, D. B., and W. D. Neigher. 1982. American psychology in the eighties: Who will buy? *American Psychologist*, 37:533-546.

Flew, A. G. N. 1980. Parapsychology: Science or pseudoscience? *Pacific Philosophical Quarterly*, 61:100-114.

Franks, F. 1981. *Polywater*. Cambridge, Mass.: MIT Press.

Furchtgott, E. 1984. Replicate, again and again. *American Psychologist*, 39:1315-1316.

Garfield, E. 1986. Refereeing and peer review: Part 1. Opinion and conjecture of the effectiveness of refereeing. *Current Contents*, 31:3-11.

Girden, E. 1978. Parapsychology: In *Handbook of Perception*, vol. 10, ed. by E. C. Carterette and M. P. Freidman. San Diego, Calif.: Academic Press.

Gray, T. 1984. University course reduces belief in the paranormal. *Skeptical Inquirer*, 8:247-251.

Gregory, R. L. 1981. *Mind in Science*. London: Penguin.

Hallam, A. 1975. Alfred Wegener and the hypothesis of continental drift. *Scientific American*, 232:88-97.

Hansel, C. E. M. 1980. *ESP and Parapsychology: A Critical Re-Evaluation*. Buffalo, N.Y.: Prometheus Books.

———. 1981. A critical analysis of H. Schmidt's psychokinesis experiments. *Skeptical Inquirer*, 5(3):26-33.

Hasted, J. B. 1976. An experimental study of the validity of metal bending phenomena. *Journal of the Society for Psychical Research*, 48:365-383.

Hebb, D. O. 1978. Personal communicaton. Cited by J. E. Alcock (1981) in *Parapsychology: Science or Magic?* Elmsford, N.Y.: Permagon.

Heskin, K. 1984. The Milwaukee project: A cautionary comment. *American Psychologist*, 39:1316-1317.

Honorton, C. 1975. Error some place! *Journal of Communication*, 25:103-116.

———. 1976. Has science developed the competence to confront claims of the paranormal? In *Research in Parapsychology 1975*, ed. by W. G. Roll, R. L. Morris, and J. D. Morris.

———. 1978a. Has science developed the competence to confront claims of the paranormal? In *The Signet Handbook of Parapsychology*, ed. by M. Ebon. New York: New American Library.

———. 1978b. Psi and internal attention states: Information retrieval in the ganzfeld. In *Psi and States of Awareness*, ed. by B. Shapiro and L. Coly. Parapsychology Foundation.

———. 1979. Methodological issues in free-response psi experiments. *Journal of the American Society for Psychical Research*, 73:381-394.

———. 1981. Beyond the reach of sense: Some comments on C. E. M. Hansel's *ESP and Parapsychology: A Critical Re-Evaluation*. Journal of the American Society for Psychical Research, 75:155-166.

———. 1985. Meta-analysis of psi ganzfeld research: A response to Hyman. *Journal of Parapsychology*, 49:51-91.

Hövelmann, G. H., and S. Krippner. 1986. Charting the future of parapsychology. *Parapsychology Review*, 17:1-5.

Hyman, R. 1977a. The case against parapsychology. *The Humanist*, 37:47-49.

———. 1977b. Psychics and scientists. "Mind-Reach" and remote viewing. *The Humanist*, 37:6-20.

———. 1981. Further comments on Schmidt's PK experiments. *Skeptical Inquirer*, 5:34-40.

———. 1985a. A critical historical overview of parapsychology. In *A Skeptic's Handbook of Parapsychology*, ed. by P. Kurtz. Buffalo, N.Y.: Prometheus Books.

———. 1985b. The ganzfeld/psi experiment: A critical appraisal. *Journal of Parapsychology*, 49:3-49.

Hyman, R., and C. Honorton. 1986. A joint communique: The psi ganzfeld controversy. *Journal of Parapsychology*, 50:351-364.

Irwin, H. J. 1985a. A study of the measurement and the correlates of paranormal belief. *Journal of the American Society for Psychical Research*, 79:301-326.

———. 1985b. Fear of psi and attitude to parapsychological research. *Parapsychology Review*, 16:1-4.

Jahn, R. G., R. D. Nelson, and B. J. Dunne. 1985. Variance effects in REG series score distributions. Paper presented at the Parapsychological Association convention, Tufts University, Medford, Mass.

Kamin, L. 1974. *The Science and Politics of IQ*. Hillsdale, N.J: Erlbaum.

Karnes, E. W., J. Ballou, E. P. Susman, and P. Swaroff. 1979. Remote viewing: Failures to replicate with control comparisons. *Psychological Reports*, 49:963-973.

Karnes, E. W., and E. P. Susman. 1979. Remote viewing: A response bias interpretation. *Psychological Reports*, 44:471-479.

Karnes, E. W., E. P. Susman, P. Klusman, and L. Turcotte. 1980. Failures to replicate remote-viewing using psychic subjects. *Zetetic Scholar*, 6:66-76.

Krippner, S., ed. 1977. *Advances in Parapsychological Research*, vol. 1, Psychokinesis. New York: Plenum.

Krippner, S., ed. 1978a. *Advances in Parapsychological Research*, vol. 2, *Extrasensory Perception*. New York: Plenum.
———. 1978b. The importance of Rosenthal's research for parapsychology. *Behavioral and Brain Sciences*, 1:398-399.
———. ed. 1982a. *Advances in Parapsychological Research*, vol. 3. New York: Plenum.
———. 1982b. Psychic healing. In *Psychical Research: A Guide to Its History, Principles, and Practices*, ed. by I. Grattan-Guinness. New York: Plenum.
———. 1983. Comments on Hövelmann's "Seven Recommendations for the Future Practice of Parapsychology. *Zetetic Scholar*, 151-153.
———, ed. 1984. *Advances in Parapsychological Research*, vol. 4. Jefferson, N.C.: McFarland.
Kurtz, P. 1986. *The Transcendental Temptation*. Buffalo, N.Y.: Prometheus Books.
Layton, B. D., and B. Turnbull. 1975. Belief, evaluation, and performance on an ESP task. *Journal of Experimental Social Psychology*, 2:166-180.
Lykken, D. T. 1968. Statistical significance in psychological research. *Psychological Bulletin*, 70:151-159.
Mabbett, I. W. 1982. Defining the paranormal. *Journal of Parapsychology*, 46:337-354.
Mackenzie, B., and S. L. Mackenzie. 1980. Whence the enchanted boundary? Sources and significance of the parapsychological tradition. *Journal of Parapsychology*, 44:125-166.
Marks, D. 1981a. Sensory cues invalidate remote viewing experiments. *Nature*, 292:177.
———. 1981b. On the review of *The Psychology of the Psychic*: A reply to Dr. Morris. *Journal of the American Society for Psychical Research*, 75:197-203.
Marks, D., and R. Kammann. 1978. Information transmission in remote viewing experiments. *Nature*, 272:680-681.
———. 1980. *The Psychology of the Psychic*. Buffalo, N.Y.: Prometheus Books.
Marks, D., and C. Scott. 1986. Remote viewing exposed. *Nature*, 319:444.
Marshall, G. D., and P. G. Zimbardo. 1979. Affective consequences of unexplained arousal. *Journal of Personality and Social Psychology*, 37:970-988.
Martin, B. 1983. Psychic origins in the future. *Parapsychology Review*, 14:1-7.
Maslach, C. 1979. Negative emotional biasing and unexplained arousal. *Journal of Personality and Social Psychology*, 37:953-969.
Mauskopf, S. H., and M B. McVaugh. 1980. *The Elusive Science*. Johns Hopkins University Press.
May, E. C., B. S. Humphrey, and G. S. Hubbard. 1980. *Electronic System Perturbation Techniques*. Stanford, Calif.: SRI International.
McClenon, J. 1982. A survey of elite scientists: Their attitudes toward ESP and parapsychology. *Journal of Parapsychology*, 46:127-152.
———. 1984. *Deviant Science*. Philadelphia: University of Pennsylvania Press.
McConnell, R. A. 1977. The resolution of conflicting beliefs about the ESP evidence. *Journal of Parapsychology*, 41:198-214.
———. 1983. *An Introduction to Parapsychology in the Context of Science*. Published by the author.
McConnell, R. A., and T. K. Clark. 1980. Training, belief, and mental conflict within the Parapsychological Association. *Journal of Parapsychology*, 44:245-268.
McGuire, G. R. 1984. The collective subconscious: Psychical research in French psychology (1880-1920). Paper presented at the 92nd annual meeting of the American Psychological Association, Toronto, August 25-28.

Miller, N. E. 1978. Biofeedback and visceral learning. *Annual Review of Psychology*, 29:373-404.

Miller, N. E., and L. Dicara. 1967. Instrumental learning of heart rate changes in curarized rats: Shaping and specificity to discriminative stimuli. *Journal of Comparative and Physiological Psychology*, 63:12-19.

Moore, L. 1977. *In Search of White Crows*. Oxford, U.K.: Oxford University Press.

Morris, R. L. 1980. Some comments on the assessment of parapsychological studies: A review of *The Psychology of the Psychic*, by David Marks and Richard Kammann. *Journal of the American Society for Psychical Research*, 74:425-443.

———. 1981. Dr. Morris replies to Dr. Marks. *Journal of the American Society for Psychical Research*, 75:203-207.

Moss, T. 1976. Uri's magic. In *The Geller Papers*, ed. by C. Panati. New York: Houghton Mifflin.

Murphy, G. 1961. *The Challenge of Psychical Research*. New York: Harper & Row.

———. 1971. The problems of repeatability in psychical research. *Journal of the American Society for Psychical Research*, 65:3-16.

Neher, A. 1980. *The Psychology of Transcendence*. Englewood Cliffs, N.J.: Prentice Hall.

Nelson, R. D., B. J. Dunne, and R. Jahn. 1984. *An REG Experiment with Large Data Base Capability, 3: Operator Related Anomalies* (Technical Note PEAR 84003). School of Engineering/Applied Science, Princeton University.

Office of Technology Assessment. 1989. Report of a workshop on experimental parapsychology. *Journal of the American Society for Psychical Research*, 83:317-339.

Osis, K., and E. Haraldsson. 1978. Deathbed observations by physicians and nurses: A crosscultural survey. In *The Signet Handbook of Parapsychology*, ed. by M. Ebon. New York: New American Library.

Oteri, L., ed. 1975. *Quantum Physics and Parapsychology*. Parapsychology Foundation.

Otis, L., and J. E. Alcock. 1982. Factors affecting extraordinary belief. *Journal of Social Psychology*, 118:77-85.

Palmer, J. 1978. Extrasensory perception: Research findings. In *Advances in Parapsychological Research*, vol. 2, ed. by S. Krippner. New York: Plenum.

———. 1983. Comments on Hövelmann's "Seven recommendations for the future practice of parapsychology." *Zetetic Scholar*, 11:164-166.

———. 1985a. Progressive skepticism: A critical approach to the psi controversy. Paper presented at the 11th annual meeting of the Society for Philosophy and Psychology, Toronto.

———. 1985b. Psi research in the 1980s. *Parapsychology Review*, 16:1-4.

———. 1985c. An evaluative report on the current status of parapsychology. Paper prepared for the United States Army Research Institute for the Behavioral and Social Sciences (final draft).

———. 1986a. Progressive skepticism: A critical approach to the psi controversy. *Journal of Parapsychology*, 50:29-42.

———. 1986b. ESP research findings: The process approach. In *Foundations of Parapsychology*, by H. L. Edge, R. L. Morris, J. H. Rush, and J. Palmer. London: Routledge & Kegan Paul.

Palmer, J. A., C. Honorton, and J. Utts. 1989. Reply to the National Research Council study on parapsychology. *Journal of the American Society for Psychical Research*, 83:31-49.

Parker, A. 1978. A holistic methodology in psi research. *Parapsychology Review*, 9:1-6.
Phillips, P. R. 1979. Some traps in dealing with our critics. *Parapsychology Review*, 10:7-8.
———. 1984. Measurement in quantum mechanics. *Journal of the Society for Psychical Research*, 52:297-306.
Pinch, T. J. 1985. Theory testing in science—The case of solar neutrinos: Do crucial experiments test theories or theorists? *Philosophy of the Social Scientists*, 15: 167-187.
Pratt, J. G. 1953. The homing problem in pigeons. *Journal of Parapsychology*, 17:34-60.
———. 1956. Testing for an ESP factor in pigeon homing. In *Extrasensory Perception*, ed. by G. E. W. Wolstenholme and E. C. P. Miller. Secaucus, N.J.: Citadel Press.
———. 1974. In search of a consistent scorer. In *New Directions in Parapsychology*, ed. by J. Beloff. London: Elek Science.
Puthoff, H. E., and R. Targ. 1974. Information transfer under conditions of sensory shielding. *Nature*, 252:602-607.
———. 1979. A perceptual channel for information transfer over kilometer distances: Historical perspective and recent research. In *Mind at Large*, ed. by C. T. Tart, H. E. Puthoff, and R. Targ. New York: Praeger.
———. 1981. Rebuttal of criticisms of remote viewing experiments. *Nature*, 292:388.
Radtke, R. C., L. L. Jacoby, and G. D. Goedel. 1971. Frequency discrimination as a function of frequency of repetition and trials. *Journal of Experimental Psychology*, 89:78-84.
Randi, J. 1975. *The Magic of Uri Geller*. New York: Random House.
Rao, K. R. 1982. Science and the legitimacy of psi. *Parapsychology Review*, 13:1-6.
———. 1983. From proof to acceptance. *Parapsychology Review*, 14:1-3.
———. 1985. The ganzfeld debate. *Journal of Parapsychology*, 49:1-2.
Reed, G. 1972. *The Psychology of Anomalous Experience*. Boston, Mass.: Houghton Mifflin. Revised edition published by Prometheus Books, Buffalo, N.Y., 1988.
Rhine, J. B. 1943. ESP, PK and the surivival hypothesis [Editorial]. *Journal of Parapsychology*, 7:223-227.
Rhine, J. B., and J. G. Pratt. 1957/1962. *Parapsychology: Frontier Science of the Mind*. Springfield, Ill.: Charles C. Thomas.
———. 1961. A reply to the Hansel critique of the Pearce-Pratt series. *Journal of Parapsychology*, 25:92-98.
Rhine, L. E. 1967. Parapsychology, then and now. *Journal of Parapsychology*, 31:231-248.
———. 1977. Research methods with spontaneous cases. In *Handbook of Parapsychology*, ed. by B. B. Wolman. New York: Van Nostrand Reinhold.
Rogo, D. S. 1986. Psi and the scientific mind: Some historical notes. *Parapsychology Review*, 17:5-8.
Roll, W. G. 1982. The changing perspective on life after death. In *Advances in Parapsychological Research*, vol. 3, ed. by S. Krippner. New York: Plenum.
Rosenthal, R., and D. B. Rubin. 1978. Interpersonal expectancy effects: The first 345 studies. *Behavioral and Brain Sciences*, 11:377-415.
Rush, J. H. 1986a. Spontaneous psi phenomena: Case studies and field investigations. In *Foundations of Parapsychology*, by H. L. Edge, R. L. Morris, J. H. Rush, and J. Palmer. London: Routledge & Kegan Paul.
———. 1986b. Parapsychology: A historical perspective. In *Foundations of Parapsychology*, by H. L. Edge, R. L. Morris, J. H. Rush, and J. Palmer. London: Routledge & Kegan Paul.

Rush, J. H. 1986c. Physical and quasi-physical theories of psi. In *Foundations of Parapsychology*, by H. L. Edge, R. L. Morris, J. H. Rush, and J. Palmer. London: Routledge & Kegan Paul.

Schachter, S., and J. E. Singer. 1979. Comments on the Maslach and Marshall–Zimbardo experiments. *Journal of Personality and Social Psychology*, 37:989-995.

Schlitz, M., and E. Gruber. 1980. Transcontinental remote viewing. *Journal of Parapsychology*, 44:305-317.

———. 1981. Transcontinental remote viewing. A rejudging. *Journal of Parapsychology*, 45:233-237.

Schlitz, M., and J. M. Haight. 1984. Remote viewing revisited: An intrasubject replication. *Journal of Parapsychology*, 48:39-49.

Schmeidler, G. R. 1971. Parapsychologists' opinions about parapsychology, 1971. *Journal of Parapsychology*, 35:208-218.

———. 1977. Methods for controlled research on ESP and PK. In *Handbook of Parapsychology*, ed. by B. B. Wolman. New York: Van Nostrand Reinhold.

———. 1984. Psychokinesis: The basic problem, research methods and findings. In *Advances in Parapsychological Research*, vol. 4, ed. by S. Krippner. Jefferson, N.C.: McFarland.

———. 1985. Belief and disbelief in psi. *Parapsychology Review*, 16:1-4.

Schmidt, H. 1969a. Precognition of a quantum process. *Journal of Parapsychology*, 33:99-108.

———. 1969b. Clairvoyance tests with a machine. *Journal of Parapsychology*, 33:300-306.

———. 1970a. A PK test with electronic equipment. *Journal of Parapsychology*, 34:175-181.

———. 1970b. A quantum mechanical random number generator for psi tests. *Journal of Parapsychology*, 34:219-224.

———. 1970c. PK research with animals as subjects. *Journal of Parapsychology*, 34:255-261.

———. 1973. PK tests with a high-speed random number generator. *Journal of Parapsychology*, 37:105-118.

———. 1974. Comparison of PK action on two different random number generators. *Journal of Parapsychology*, 38:47-55.

———. 1975. Toward a mathematical theory of psi. *Journal of the American Society for Psychical Research*, 69:267-292.

———. 1976. PK effect on pre-recorded targets. *Journal of the American Society for Psychical Research*, 70:267-291.

———. 1978. A take-home test in PK with pre-recorded targets. In *Research in Parapsychology 1977*, ed. by W. G. Roll. Metuchen, N.J.: Scarecrow Press.

———. 1979a. Search for psi fluctuations in a PK test with cockroaches. In *Research in Parapsychology 1978*, ed. by W. G. Roll. Metuchen, N.J.: Scarecrow Press.

———. 1979b. Use of stroboscopic light as rewarding feedback in a PK test with pre-recorded and momentarily-generated random events. In *Research in Parapsychology 1978*, ed. by W. G. Roll. Metuchen, N.J.: Scarecrow Press.

———. 1979c. Evidence for direct interaction between the human mind and external quantum processes. In *Mind at Large*, ed. by C. T. Tart, H. E. Puthoff, and R. Targ. New York: Praeger.

———. 1981. PK tests with pre-recorded and pre-inspected seed numbers. *Journal of Parapsychology*, 45:87-98.

———. 1985. Addition effect for PK on prerecorded targets. *Journal of Parapsychology*, 49:29-244.

Schmidt, H., R. L. Morris, and L. Rudolph. 1986. Channeling evidence for a PK effect to independent observers. *Journal of Parapsychology*, 50:1-15.

Schmidt, H., and L. Pantas. 1972. Psi tests with internally different machines. *Journal of Parapsychology*, 36:222-232.

Scott, C. 1985. Why parapsychology demands a critical response. In *A Skeptic's Handbook of Parapsychology*, ed. by Paul Kurtz. Buffalo, N.Y.: Prometheus Books.

Scott, C., and K. M. Goldney, 1960. The Jones boys and the ultrasonic whistle. *Journal of the Society for Psychical Research*, 40:249-260.

Seligman, M. E. P., and J. L. Hager. 1972. The sauce-Bernaise syndrome. *Psychology Today*, 6:59-61.

Sheils, D., and P. Berg. 1977. A research note on sociological variables related to belief in psychic phenomena. *Wisconsin Sociologist*, 14:24-31.

Sommer, R., and B. A. Sommer. 1983. Mystery in Milwaukee: Early intervention, IQ, and psychology textbooks. *American Psychologist*, 38:982-985.

———. 1984. Reply from Sommer and Sommer. *American Psychologist*, 39:1318-1319.

Spanos, N. P. 1986. Hypnotic behavior: A social psychological interpretation of amnesia, analgesia, and "trance logic." *Behavioral and Brain Sciences*, 9:449-467.

Stanford, R. G. 1977. Experimental psychokinesis: A review from diverse perspectives. In *Handbook of Parapsychology*, ed. by B. B. Wolman. New York: Van Nostrand Reinhold.

———. 1978. Education in parapsychology: An overview. *Parapsychology Review*, 9:1-12.

Sterling, T. C. 1959. Publication decisions and their possible effects on inferences drawn from tests of significance—or vice versa. *Journal of the American Statistical Association*, 54:30-34.

Stevenson, I. 1977. Reincarnation: Field studies and theoretical issues. In *Handbook of Parapsychology*, ed. by B. B. Wolman. New York: Van Nostrand Reinhold.

Swets, J. A. 1988. Preface. *Enhancing Human Performance*, pp. vii-xi. Washington, D.C.: National Academy Press.

Targ, R., and K. Harary. 1984. *The Mind Race*. New York: Random House.

Targ, R., and H. E. Puthoff. 1974. Information transfer under conditions of sensory shielding. *Nature*, 251:602-607.

———. 1977. *Mind-Reach*. New York: Delacorte.

Targ, R., H. E. Puthoff, and E. C. May. 1979. Direct perception of remote geographical locations. In *Mind at Large*, ed. by C. T. Tart, H. E. Puthoff, and R. Targ. New York: Praeger.

Tart, C. T. 1980. Comments on Karnes et al. *Zetetic Scholar*, 6:85-86.

———. 1982. The controversy about psi: Two psychological theories. *Journal of Parapsychology*, 46:313-320.

———. 1984. Acknowledging and dealing with the fear of psi. *Journal of the American Society for Psychical Research*, 78:133-143.

Tart, C. T., H. E. Puthoff, and R. Targ. 1980. Information transmission in remote viewing experiments. *Nature*, 284:191.

Taylor, J. G. 1975. *Superminds*. New York: Viking.

Taylor, J. G., and E. Balanovski. 1979. Is there any scientific explanation of the paranormal? *Nature*, 279:631-633.

Terry, J., and H. Schmidt. 1978. Conscious and unconscious PK tests with prerecorded targets. In *Research in Parapsychology 1977*, ed. by W. G. Roll. Metuchen, N.J.: Scarecrow Press.

Thalbourne, M. A. 1982. *A Glossary of Terms Used in Parapsychology*. London: Heinemann.

———. 1984. The conceptual framework of parapsychology: Time for a reformation. Paper presented at the 27th annual meeting of the Parapsychological Association, Dallas.

Thouless, R. H. 1942. The present position of experimental research into telepathy and related phenomena. *Proceedings of the Society for Psychical Research*, 42:1-19.

Truzzi, M. 1982. Editorial. *Zetetic Scholar*, 10:5-6.

———. 1985. Anomalistic psychology and parapsychology: Conflict or detente? Paper presented at the annual meeting of the American Psychological Association, Los Angeles, August 23.

Tyler, L. E. 1981. More stately mansions—Psychology extends its boundaries. *Annual Review of Psychology*, 32:1-20.

Wagner, M. H., and M. Monnet. 1979. Attitudes of college professors towards extrasensory perception. *Zetetic Scholar*, 5: 7-16.

Walker, E. H. 1974. Consciousness and quantum theory. In *Psychic Exploration*, ed. by J. White. New York: Putnam.

———. 1975. Foundations of paraphysical and parapsychological phenomena. In *Quantum Physics and Parapsychology*, ed. by L. Oteri. Parapsychology Foundation.

———. 1984. Introduction. *Advances in Parapsychological Research*, vol. 4, ed. by S. Krippner. Jefferson, N.C.: McFarland.

Weiner, D. H., and N. L. Zingrone. 1986. The checker effect revisited. *Journal of Parapsychology*, 50:85-121.

Wolman, B. B., ed. 1977a. *Handbook of Parapsychology*. New York: Van Nostrand Reinhold.

———. 1977b. Mind and body: A contribution to a theory of parapsychological phenomena. In *Handbook of Parapsychology*, ed. by B. B. Wolman. New York: Van Nostrand Reinhold.

Yalow, R. S. 1978. Radioimmunoassay: A probe for the fine structure of biologic systems. *Science*, 19:1236-1245.

Zusne, L., and W. H. Jones. 1982. *Anomalistic Psychology: A Study of Extraordinary Phenomena of Behavior and Experience*. Hillsdale, N.J.: Erlbaum.

DATE DUE

FEB 03 1994			
MAR 1 8 1995			

GAYLORD   PRINTED IN U.S.A.